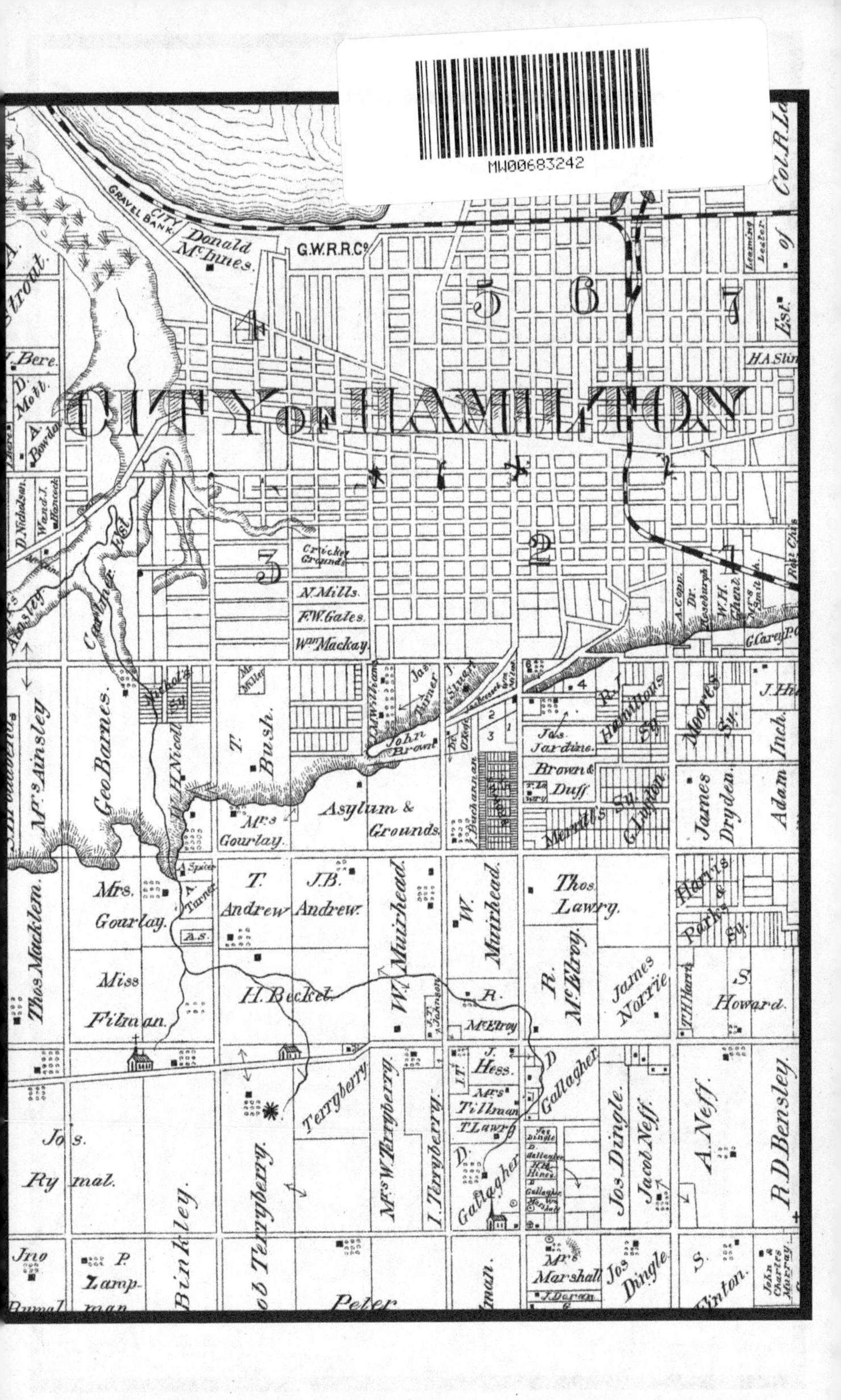

MW00683242
CITY OF HAMILTON
G.W.R.R.Co.
CITY GRAVEL BANK
Donald McInnes.
Cricket Grounds
N. Mills.
F.W. Gates.
Wm. Mackay.
Asylum & Grounds.
John Brown
Jos. Jardine.
Brown & Duff.
Hamilton's Sq.
Moore's Sq.
James Dryden.
Adam Inch.
Geo Barnes.
Mrs Ainsley
T. Bush.
Mrs Gourlay.
Mrs. Gourlay.
T. Andrew
J.B. Andrew.
W. Muirhead.
Thos Lawry.
Thos Macklen.
Miss Filman.
H. Becket.
R. McElroy.
James Norrie
S. Howard.
Terryberry.
Mrs W. Terryberry.
I. Terryberry.
J. Hess.
Mrs Tillman
T. Lawry
D. Gallagher
Jos. Dingle.
Jacob Neff.
A. Neff.
R D Bensley.
Jos. Rymal.
Binkley.
P. Zamp-man
Peter
Mrs Marshall
Jos Dingle.
S. Finton.
John & Charles Murray.

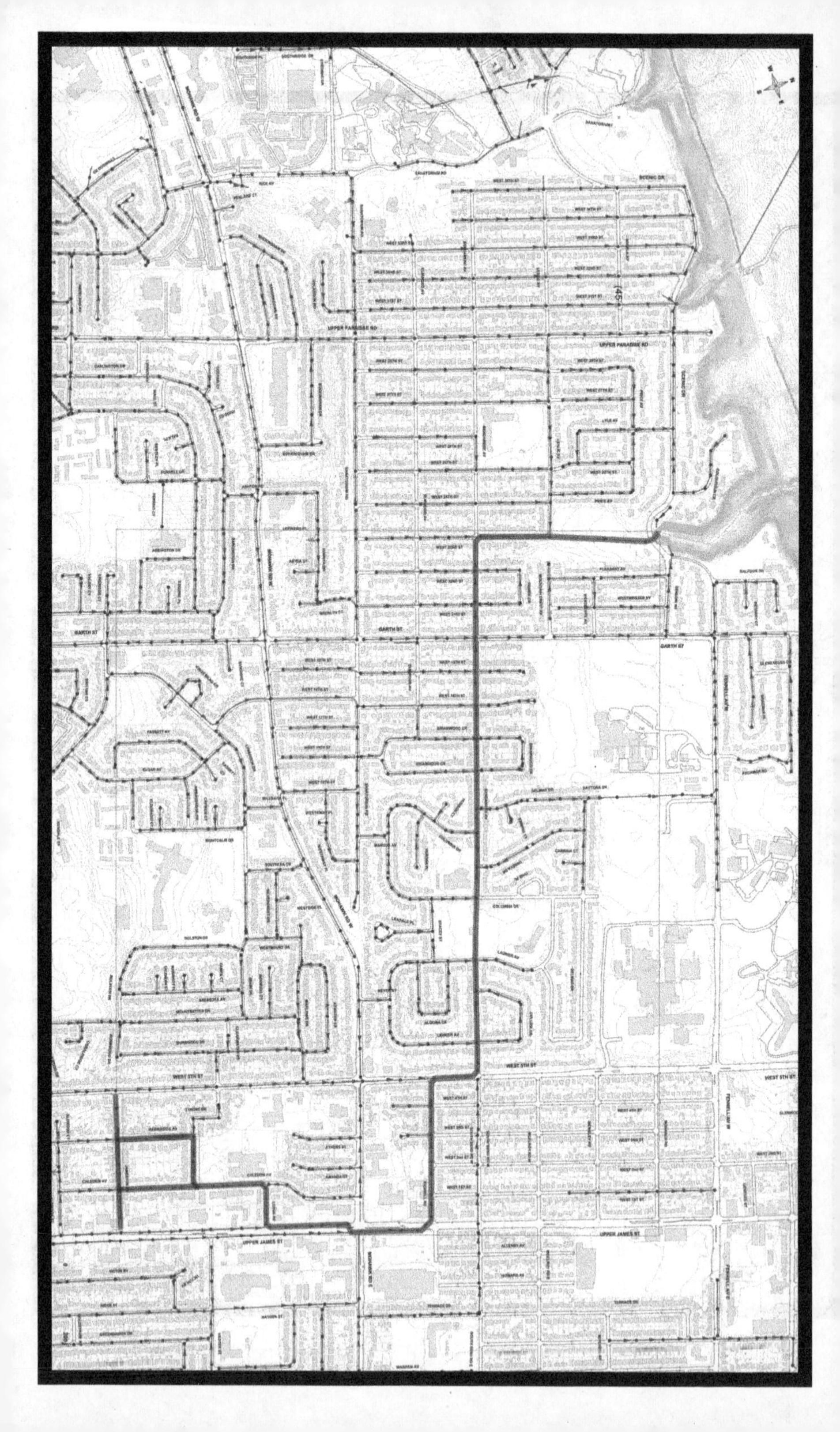

DAYLIGHTING CHEDOKE

Other Works by John Terpstra

Non-Fiction

The House With the Parapet Wall

Skin Boat: Acts of Faith and Other Navigations

The Boys, or, Waiting for the Electrician's Daughter

Falling Into Place

Poetry

Mischief

In the Company of All

Brilliant Falls

Naked Trees

Two or Three Guitars: Selected Poems

Disarmament

Devil's Punch Bowl

The Church Not Made With Hands

Captain Kintail

Forty Days & Forty Nights

Scrabbling for Repose

DAYLIGHTING CHEDOKE

Exploring Hamilton's Hidden Creek

By John Terpstra

James Street North Books is an imprint of Wolsak and Wynn Publishers.

Cover design: Dave Kuruc
Interior design: Mary Bowness
Author photograph: Jeff Tessier
Typeset in Minion
Printed by Ball Media, Brantford, Canada

Epigraph on page vii from *My Conversations with Canadians* by Lee Maracle. Used with permission

Map images by Dave Kuruc, adapted from "Barton, County of Wentworth Map, 1875" and "Storm Flow from Jameston Avenue to Chedoke Creek at Niagara Escarpment" courtesy of the City of Hamilton.

Canadä

The publisher gratefully acknowledges the support of the Canada Council for the Arts, the Ontario Arts Council and the Canada Book Fund.

Wolsak and Wynn
280 James Street North
Hamilton, ON
Canada L8R 2L3

Library and Archives Canada Cataloguing in Publication
Terpstra, John, author
Daylighting Chedoke : exploring Hamilton's hidden creek / John Terpstra.

ISBN 978-1-928088-72-1 (softcover)

1. Chedoke Creek (Ont.)–History. 2. Rivers–Ontario–Hamilton.
3. Urbanization–Ontario–Hamilton. 4. Human ecology–Ontario–Hamilton.
5. Hamilton (Ont.)–Environmental conditions. I. Title.

FC3098.56.T47 2018 551.48'30971352 C2018-903870-5

Contents

Some of our people wish Canadians would move back to their original homelands. Not me – I hope they fall in love with the land the way I have: fully, responsibly, and committed for life.

Lee Maracle,
My Conversations with Canadians

Citizen Geography

During one of our stays in New Jersey to visit her grandmother, our daughter Anna, who was around ten years old, spent a good portion of each day at Goffle Brook playing in the same water and woods where her mother and her three long-departed uncles, and also her grandmother, had played in their childhoods.

The brook formed the property line of the side yard of the house, which stood in one of the many contiguous towns that lay within an hour's commute of New York City. A short, steep slope led down to the water's edge. A path, carved and compacted into the side of the slope, followed the course of the brook behind a few neighbours' homes, made a turn and then reappeared at the end of the dead-end street around the corner from the house.

The water was shallow and the brook could easily be waded. It had a sandy bottom and small, round stones. Trees overhung its singsong. Idyllic, in short.

Anna spent most of her afternoons absorbed in creek activities while we attended to our adult errands and interests. One day we sat outside on the concrete steps of the breezeway waiting for her to answer our call home for dinner. She walked up the driveway at last, tired and happy.

"I can't get enough of this place," she said.

My first experience of freely flowing water was as a child growing up beside the North Saskatchewan River in Edmonton, Alberta, though it was not the river that captured my imagination.

Mid-winter chinooks and early spring weather produced streams that sent my younger brother and me to the basement to borrow a clothespin from the line where our mother hung laundry to dry in the winter. They were two-piece, wooden clothespins with a metal spring connecting them. A determined twist separated the pieces and provided our boats, which we would take outside to the street to race in the rivers of melting snow that ran beside the curb. Those irresistible rivers could be an obstacle course as they skirted mounds of snow and slipped under shelves of thin ice. Sometimes our boats came out the other side, somewhere ahead, sometimes they didn't. "Where are my clothespins disappearing to?" Mom shouted from the basement. We didn't know. If our boats did manage to reach the finish line, the trick was to grab them before they were swept between the bars of the storm drain, or we would have to sneak down to borrow another pin.

I hated it when my father's job transferred him east to Hamilton, Ontario. I hated with a passion moving away.

What is it about running water?

I linger indefinitely on the bridge over the narrow feeder to Borer's Creek and watch the rocks, moss and tree roots at their scenic best guide a leaping, clear band of wetness that rides over and through. How do you even describe water? The sight and sound. The touch. A two-year-old grandson stands on a step-stool in front of the kitchen sink and lets the tap water flow over and through his hands and fingers, in thrall.

I've come to see these trickling ribbons of H_2O as living things. I've come to have a bird's-eye view of these lithesome, moving waterlines that wind over the landscape, reflecting light. These open arteries. The earth's equivalent to the ones that run through our own bodies.

Am I following the creek, or is the creek following me?

At the close of a presentation on the geography and history of Hamilton based on my book *Falling Into Place* I mention Chedoke (*sh*-dohk) Creek. I no longer hate with a passion my uprooting or where our family moved to. Fifty-plus years after it happened, I can't seem to get enough of the place. My presentation had focused mainly on the older half of the city below the Niagara Escarpment and its dominant, though subtle, geographical feature, the Iroquois Bar. I wanted to give the audience on the Mountain, as we call it, something more local. The Mountain is a natural second tier to the city, a plateau one hundred metres above where the city began. Upper and lower city are two different geological epochs, and the two often seem like two different worlds.

Chedoke came to mind only because I have always liked the name and had once noticed the creek, I tell them, on a poster of an old map. The creek was a wiggly line that ran across the Mountain, fell over the escarpment and emptied into Cootes Paradise. You can't see the creek in the Mountain suburbs it winds through any longer because it's underground, buried in storm drain pipes, but I remembered from the map that it started, or had its source, somewhere near the intersection of Mohawk Road and Upper James Street, more or less in the middle of the parking lot of the Canadian Tire store. A very un-creek-like setting.

This was the extent of my Mountain content for the evening.

During the question-and-answer period that followed, a man stood and said that he was a retired builder and that when he was constructing houses on West 18th Street in the 1960s, the city came along, laid pipes and buried a creek that was running through the neighbourhood.

Given the location, that would have been Chedoke Creek, I responded, surprised at the coincidence.

Then another person in that grand audience of eight stood and told us that she knew exactly where the source of Chedoke Creek was because her grandparents owned the farm where the spring was located. It was called Spring Farm.

Suburban development on the Mountain began in earnest in the 1950s and '60s from the escarpment edge southward. Until then it was mostly farmland. During the course of development, creeks were routinely buried underground as the main channels of a storm drain system to catch road and roof runoff.

In a city, the built geography often overwhelms the natural. You'd never guess today that there are six creeks in the Chedoke watershed, if you even knew that a watershed existed. You'd never guess that these six creeks cover the entire western half of the Mountain suburbs, sending their water to six distinct falls that drop over the edge of the escarpment along a two-kilometre stretch that coincides with the path the 403 highway has chiselled into the face of the escarpment. The creeks then join together as one family in a single creek that flows through a valley that you'd also never guess exists because the highway was built through it during those same decades and now dominates the valley. Thousands of us daily drive upstream and downstream beside a creek that is squeezed between its own valley walls and six lanes of traffic, unaware.

We have no idea where we are.

I say this from personal experience, having uncovered these facts only after the presentation at Terryberry Library.

Was it the word or the water?

Somewhere during these fifty-plus years of living here, Chedoke meandered into my mindscape, nosing along, unnoticed until the day that I needed it for a presentation.

Spring Farm and the unlikely coincidence of its mention that evening prompted me to follow up on this mystery. As I did, stories began to spring and flow, a new and unexplored landscape opened up, and before I knew it, the small, wooden craft of my childhood, in the form of a pencil, was caught in the current of a brave little stream.

Place-Thought is a concept that originates in Indigenous people's experi-ence of life and the earth. It gives agency to the non-human world, as well as to the human. The term is used in an essay by Vanessa Watts, an Anishinaabe and Haudenosaunee professor and the Academic Director of Indigenous Studies at McMaster University in Hamilton.

She writes, "the land is alive and thinking."

The idea is that society is built in conversation with the world that surrounds it. In the interaction between people and place. We make pacts and historical agreements with the earth, the animal world, the sky world, the spirit world, the world of rocks and trees and flowing streams. It is an active thing. A two-way street.

Social relations, personal relations.

The connection is a sacred one. This connection, and this way of looking at things, doesn't come naturally, or easily, to someone raised in the Western tradition with an Enlightenment frame of mind.

It's the word, not the water, that most people in Hamilton recognize.

Chedoke is a common entry in the civic vocabulary. It's the name of a hospital, a school, a highway, a skating rink, a florist and a room in the convention centre downtown, to list a few. Spell-check immediately assumes that you mean to write *Cherokee*, and part of the word's attraction is certainly its Indigenous look and sound. Most people assume it is an Indigenous word. It feels original to the place, as though it belongs here, in a way that any immigrant-settler kid wants to belong. And on one level, we're all immigrant-settlers.

Chedoke is not a First Nation's word, however. Neither is it English or French, the two other most likely candidates. Nor is it Scottish or Irish, the two largest immigrant groups to come this way in the early nineteenth century. No one has yet been able to follow *Chedoke* to its source. It's unique. Indigenous to this city.

Educated guesses have been made.

"This is really not an Indian word, but a corruption of two English words, 'Seven Oaks.' The Indians caught the sounds and changed 'Seven Oaks' into Chedoke." So writes William Francis Moore in 1930 in his *Indian Place Names in the Province of Ontario.*

This explanation is as good as any other. But why *corruption*?

Chedoke is a *word-gift*.

The landscape changes when you pursue the stories connected to it. When you pursue the historical threads, the oral and written creeks that travel through time.

I was talking to Tys Theysmeyer at a presentation of the Cootes to Escarpment EcoPark System, an initiative to create an urban park in our heavily populated area of the country. He works for the Royal Botanical Gardens, which is a participant in the initiative, and some of whose lands are included in the park. Tys has been doing historical research lately into the area under discussion that evening, both on the ground and in archives. He was delving into events from the late eighteenth century, reading surveyor's journals and searching out landmarks, charting the paths of the earliest roads and how they line up with the current roads. He was talking to homeowners whose families have lived in the area for generations, some since the settlement period.

He shook his head as he said, "The place looks and feels so different to me now. It's transformed. Layered. Alive." It seemed as though he was waking from a dream. Or not from but *into* a dream, one in which the world is more than what you see before you.

We choose the landscapes we live in. Forget or ignore the stories, neglect to enter or hold up your side of the conversation, and all you are left with is what is in front of your eyes. And that will likely be a highly engineered environment made out of concrete.

Michael Hess enters British territory west of the Niagara River with his son Peter. It is 1788. Michael is almost fifty years old, while Peter is only ten. They are scouting for land, searching for a new home for their own and three other families. They don't necessarily want to leave their farms in Pennsylvania, but life has gotten a little too tense for them in those newly united States.

Father and son dock their canoe at the portage terminus at Head-of-the-Lake in the far western corner of a bay at the far western end of Lake Ontario. This portage would take them to Lake Huron if they let it. It's a major route, an ancient foot-highway that shows up on the earliest European maps of North America. It leads overland from Head-of-the-Lake to the Grand River, in what is now Brantford, where a short upriver paddle brings the traveller to a portage that carries on to the Thames River, which drains into Lake St. Clair, directly below Lake Huron. The route was in existence long before any Europeans arrived, part of an intercontinental Indigenous network of paths as well travelled as our highway systems today.

Michael and Peter climb the steep twenty-metre slope before them, survey the marshland that spreads below on the other side and continue down the path that leads into the shallow valley. There they encounter a creek. Instead of crossing the creek and staying on the portage highway, they take an unmarked exit and enter the woods to follow the creek upstream.

Ever since I learned this story from Gail Dawson, who spoke up at my library presentation, and her cousin Marlene Gallagher, both of whose grandparents lived at Spring Farm, I have been seeing the backs of father and son as they walk. His rounded shoulders, stocky and stooped forward slightly, not from age or infirmity but because he and Peter, who is his miniature, are walking up a grassy incline. He wears a black hat and carries a walking stick. They are strangely without gear. Before them rises the Niagara Escarpment, a fifty-metre-high wall that the creek plunges over, and that they will need to scale in order to pursue their goal. Their goal is to follow the creek to its source. They are looking to find arable land with the reliable source of water that the creek's spring could provide. The slope they climb will be a golf course in a little over one hundred years. The two could be golfers climbing up to the famous seventeenth tee set high into the base of the escarpment. I could be their caddy.

But I have an advantage over Michael and Peter. I know where Chedoke Creek is going, or rather, where following it upstream will take them. I know its path. I have a map. I have several more maps now, drawn over the past two hundred years. There is no map for them, though they have been given fair directions into this landscape and the network of well-trod paths and ancient highways makes it difficult to get truly lost. But in following the creek they are charting their own path.

Father and son have an advantage over me because their Chedoke Creek is flowing full tilt, undiminished by time and settlement, in open daylight from source to mouth. The sun shines on it, the moon rises over it and the rain falls in it. They walk through a countryside of dense wood, ravine, meadow and wetland. To follow them, I walk paved streets and concrete culverts and channels, enter the cave mouths of storm drains, wander housing developments, climb fences and try not to trespass too wantonly. I peer down streets from the inside of a car, seeking the creek's shallow valley path in the swells and sloped acreage of parking lots.

The creek they follow owns the countryside it flows

through. It is a liquid line that traces the surface of the earth and makes sense of the hills and swales. The landscape today makes a different kind of sense.

Michael and Peter track the creek to its source. They like what they see, return home to Pennsylvania to gather their family of ten, then come back the next year to begin clearing the land and building a new life. Michael will immediately petition the British government for ownership of the land. It will be granted. This is the whole idea for the British. If you give it away for free, they will come. If you offer two hundred acres to each married couple, plus another fifty acres for each child, then the unhappy, fearful, non-partisan, dispossessed and disenfranchised folk from the south will relocate to this place.

They will populate this wilderness.

Wilderness? Head-of-the-Lake was not unpopulated at the time. It was not empty of humans or animals, though there were very few Europeans at the far western end of Lake Ontario who were not British military or connected to the military, and the Indigenous population itself, for various historical reasons, was also small. The land lay between the traditional territories of Algonkian people to the north and Iroquoian to the south, and had become something of a no man's land.

As the Hess family and many others from the States fled, escaped or accepted the invitation and migrated north across the Niagara River, the British made treaties with the Mississaugas and other Indigenous peoples in the area. The Six Nations of the Finger Lakes area in New York State, routed from their ancestral home by the Continental Army and likewise fleeing the States, were given, for their loyalty to the British crown, ten kilometres of land on either side of the Grand River, from its mouth at Lake Erie to its source. The Six Nations were also known as the Iroquois Confederacy. Today they refer to themselves as Haudenosaunee. The People of the Longhouse.

Leading all this human traffic was the land surveyor for the Crown. He drew the lines that parcelled out the landscape in square and rectangular blocks, casting a net that drew in and trapped every living creature. Some of these settler-newcomers became very adept in the buying and selling of the parcels they had been granted, at pulling and twisting the net strings. They grew wealthy. Others, less adept, became land-rich but cash-poor and went bankrupt. Yet others neither sold nor bought, but settled down to raise generations of family on their new lands.

Meanwhile, those for whom this parcelling out was no way to conduct a conversation with a living landscape, like the Haudenosaunee, were at a real disadvantage. Their lands soon began to be whittled away in deals they had little knowledge of or control over. Today the Six Nations Reserve is two hundred square kilometres in size, or about five percent of the original grant.

Does it make any difference if the story is not factual but merely a family tale? If forty-eight-year-old Michael Hess did not, in fact, arrive at Head-of-the-Lake with his ten-year-old son, or follow Chedoke Creek from its mouth to its source at a spring three kilometres south of the escarpment brow?

It turns out that Michael Hess travelled with the heads of the three other families, the Kribs, Smiths and Rymals, who were looking to relocate here from Pennsylvania. All were second-generation German, with Anglicized names (Hesse, Krebs, Schmidt, Reimel), whose parents had emigrated from Europe and remained loyal to the British for having provided them a North

American haven. Michael's great (x4) grandson John Gallagher wrote a family history in the 1980s. Based on his archival research, he states categorically that this group of four men travelled overland rather than by canoe, and set up camp at what is now the corner of Upper James and Mohawk Road. Both of these were well-worn paths in the Indigenous network: Upper James as a portage that led south from the bay below the escarpment to the Grand River, and the Mohawk Trail as a major route across Upper New York State and the Niagara Peninsula.

Michael Hess & Co. camped on a low hill overlooking a creek at the northwest corner of the intersection where a grocery store and a strip mall now stand. The land had not yet been officially surveyed, but from there the men fanned out and spent a couple of days scouting out and choosing where to establish new farms to which they could bring their families.

From the camp, Michael followed the creek upstream to its source, one half-kilometre south, in the shadow of an outcrop of rock, and chose to settle there.

Everything becomes discovery. Or un-covery. New territory. Every walk and conversation, and each visit to the library, where I stare yet again at the large, hanging city maps from the 1920s.

I know that Chedoke Creek below the Mountain flows through a big, concrete pipe called the Chedoke Storm Drain under Chedoke Civic Golf Club. I thought the shallow creek that runs in a ravine parallel to the buried pipe, between the golf course and Chedoke Avenue (the names keep coming), was overflow, or some kind of small tributary. Frankly, I gave it very little thought at all. Now I see that the pipe wasn't yet laid and didn't exist in the 1920s, and the wooded vale between the fairway and the homes shades the original creek itself.

So I went to explore. Time to get my feet wet.

A trail leads upstream from Glenside Avenue, a one-house stub of a street that runs off Chedoke Avenue into the golf course. The trail winds through a narrow woodlot that borders the creek. The water is not always visible because the path follows the top of the ravine. Property lines run across the creek, and some homeowners have brought deck chairs down and created little garden oases on one side or the other. Some, not many, have erected chain-link fencing to enclose their portion of what otherwise feels like a public benefit.

A short way upstream, the path leads out of the woods and onto the golf course. This is the sloped lawn I had imagined Michael and Peter to be walking, with me tagging behind. Without the golf course having cleared the landscape, theirs would have been a tougher slog than I imagined. A little farther along, a path leads back into the woods. It stops dead after a few metres. Strange. I retreat back onto the fairway, then take the next path into the trees. It also stops dead. Then I find a golf ball and it dawns on me: these stubs of paths are golfers' entry points into the overgrown rough that has swallowed their hook or slice.

Nearer the escarpment, the ravine deepens. Getting down to the water's edge proves tricky – as does keeping my shoes dry once I manage to reach the mud flat there. My feet are officially wet. The sound of falling water comes from the creek pouring over a log that looks placed, not fallen. Behind the log, a small pool bubbles up like a spring. It is a spring, though not a natural one. The storm drain channels ninety percent of the creek through its concrete pipe, but is rigged with a smaller pipe that provides the original, parallel creek this false spring, which guarantees a continual, controlled flow, one that never floods. Creek as neighbourhood garden feature.

It's a lonely, forlorn and forgotten spot down there by the bubbling pond.

What looks like a thin, fallen branch is sticking out of the wild flowers nearby, though it's too straight for any tree branch. I reach in and pull out a putter. Left-handed. Signed, Arnold Palmer. Flung from the fifteenth green by no Arnold Palmer.

A clue to the source of the word *Chedoke* is the twenty-five-room stone house that stands on the escarpment brow just above the spot where I stood pondering the putter.

"This is the original Chedoke," claimed Wilson Balfour-Baxter, who was living in the house when she and other Balfour family heirs deeded it to the city in 1977. The Balfours had lived there since the turn of the century. Wilson continued to live there after the gift was made, paying rent and keeping up with the utilities, until she passed away in 2013 at the age of ninety-seven. Then the city got the keys.

William Scott Burn built the house in 1836 using stone from the escarpment. Maybe those seven oaks were on his property. He didn't live there long. "The House and Grounds of 'Chedoke'" was sold at auction in 1842. It's the first recorded use of the name that I have come across. So maybe Wilson is right. The grounds included the creek and falls, which by the latter half of the nineteenth century had already made a name for itself on postcards of local scenic sites.

Among Chedoke House's many listed features was "a pump in the Kitchen, with a never-failing supply of excellent water."

What a morning. The Spring Farm cousins, Gail Dawson and Marlene Gallagher-Gravefell ("I married a Gravefell, but I'm taking Gallagher back."), gave me a guided tour.

. . . a stream of very cold, clear water . . ., a crystal clear spring . . ., a spring of pure gushing water . . .

These are descriptions in three different tellings of the story of how Michael Hess came to choose this particular location for his new family farm. The journals and typewritten pages and news clippings lie spread out on Marlene's dining-room table. Michael chose land that became Lot 14 on the Sixth Concession. It was a rectangular parcel bounded by the Indigenous paths that are now Mohawk Road and Upper James Street on the north and east, together with what are now West 5th Street and Limeridge Road on the west and south. This remained the core of the family farm. The Hess family name remained with the property until one of Michael's great-grand-daughters married Dan Gallagher in the 1860s, and the married couple took over Spring Farm. It stayed in the second family's possession well into the twentieth century.

After poring over the books and binders for an hour or so, we launched into a tour of the block bounded by those four streets in Gail's car, myself in the back seat, craning to see as they point out the homes and locations where their uncles once lived, their aunts, great-uncles, grandparents and great-grand-parents, the house where Marlene grew up, where the Gallagher Brothers lime kiln business was located, and the row of houses built for employees. The block was populated by generations of family members. Some of their homes still stand, some not. A great-uncle's home still stands off Lotus Avenue, partially hidden behind a newer structure. Another great-uncle's home, an Edwardian two-and-a-half-storey brick, fronts onto Upper James and houses a veterinary clinic. The only remnant of the lime kiln business is a row of workers' homes and Quarry Court, where Marlene points out the bowl-shaped depression with houses standing inside its semicircle.

The cousins fill in and correct each other's memories, Marlene from the perspective of having lived on Lot 14 (though her mailing address on West 5th Street was Lot 8, Concession 6, she says), Gail from the perspective of someone who visited as a child this large neighbourhood of extended family that her mother missed so dearly. Their family lived in Toronto. We are touring an area that has been transformed, in two generations, from the farmland, orchard and largely abandoned quarry they knew as children into a landscape of low-slung homes and buildings, wide pavements and the continuous flow of automobiles. Their eyes help mine to see beyond the suburban and commercial spread Spring Farm has become, and almost redeems this dismal landscape. But nothing can rescue the spring itself. Where was it? Where *is* it? Marlene pointed to where she thinks the water once bubbled up out of the ground, filled a small pond and started a creek. As near as she could tell, it lies buried midway under Jameston Avenue, which stretches between Upper James and West 5th.

In their time, the creek flowed through a culvert under Upper James Street, where it widened and, in winter, froze. Local children skated there, where an arena was built in the 1960s. My family moved to Hamilton around that time, and I played hockey in that newly built Mountain Arena, which is now named after a local hockey hero, Dave Andreychuk. I walked, on winter mornings, from our new suburb to hockey practice at 5:30 a.m. through the undeveloped countryside between West 5th and Upper James. Across Spring Farm, I now know. I can almost remember walking past a line of trees beside a frozen creek, but the memory may be wishful.

Marlene remembered skating long distances on the frozen

creek, and says that as kids, her father and Gail's mother, who were brother and sister, once came upon a young Indigenous girl while they were skating. A brief sighting of a young Haudenosaunee who has since taken up residence in the family story. Is she an escapee from the surveyor's net or someone for whom the net does not apply, who is at home in this landscape wherever she is?

Time is not so linear to me any longer. This is partly an age thing. It does not run, like a creek, as it seemed to when I was younger. Actually, on a week-to-week basis, it does run like a creek, a gushing creek. Here and gone.

Yet it lingers and pools too. And when it does, what happened yesterday and what happened sixty or two hundred years ago feel concurrent. You can play along its edges. You can see to the bottom. Or what looks like the bottom.

This does not mean that you can see clearly.

Our infrastructure expedition has been a few months in the planning, and the weather is perfect. Cloudless and bright.

Beyond the sign that reads "Authorized Parking Only," the 403 sweeps noisily through the valley below, with the creek channelled in concrete beside it. We have parked at the far corner of a large lot on the opposite side of Longwood Road from the old Westinghouse headquarters, which has since become McMaster Innovation Park. Mary and I decide that the fact that we are a sixty-year-old couple who are going on an outing to explore the Chedoke Creek Storm Drain with their twentysomething nephew Nigel and his girlfriend, Dorna, is authorization enough.

The four of us gingerly descend through the weeds and wild

flowers of the steep valley slope. It's a surprising, visually rich scene down there. Big slabs of concrete, cracked and broken, lie dislodged and angled like shifting tectonic plates. That's Mary's image: tectonic plates. It's so apt she deserves citing. The world of the engineered watercourse is overgrown and lush with greenery. A tree trunk fills the gap where a slab has gone missing. Another grows on top of the channel side. Reeds and weeds populate all cracks, which are many. A steady stream of clear, shallow water covers the concrete creek bed from one side to the other, while the sound of bigger water, loud and rushing, competing with traffic, echoes from inside the storm drain where we are headed. We are glad for our rubber boots. And flashlights.

And so we enter the gaping mouth of concrete that is the outfall end of the storm drain, and begin to walk up the tongue of running water.

You could drive a car through here. We take photos fore and aft as the bright window of sunlight behind us diminishes to the size of a pinhole and the world inside dims to solid darkness. The rushing sound gets louder and louder, and we soon know why. The creek cascades down a long stairway, a series of twenty steps with two-metre-deep treads. This has been dubbed the Stairway to Paradise by Michael Cook, the author of a website called Vanishing Point. The site is dedicated to what he calls "citizen geography," the rivers and streams in the largely invisible, subterranean world of storm drain infrastructure. A world that our urban lives depend upon.

We march up and on, examining the crazily stained walls as we go, the many pencil-thin stalactites that hang from cracks in the ceiling, and the ends of drainpipes of various diameters that have their openings halfway up the wall and are adding their trickling contributions to the creek. Occasionally, a row of ladder rungs built into the concrete walls leads up through square shafts to manhole covers that are letting in parcels of natural light. We feel truly down among roots now, and have a warm-blooded urge to reach for that light. Nigel climbs up through the spiderwebs, and the spiders, and tries to use the GPS on his cellphone to locate us. Are we under Aberdeen Avenue or the railway tracks? The signal is lost by the time he comes down, so we are none the wiser. Being underground and in total darkness is the definition of disorientation.

Nigel describes this square tunnel as a cut-and-cover construction where a trench is dug, the tunnel formed-up, concrete poured and the finished result is buried. For a time, we find ourselves inside a concrete cylinder, a big tube wide enough in diameter that we can still walk upright, alternately straddling the current or splashing our way through. For this round stretch of tunnel, he says, a trench would have been dug and precast sections of the pipe laid, prior to the trench being filled in again. The current here is strong and deep enough to occasionally wash over boots that come halfway to our knees. Overall, the concrete of both the cut-and-cover and conduit is in remarkably good condition for being more than fifty years old, though there are places, mostly at section seams, where the concrete is crumbled or broken and we have to watch for a pitfall.

The tunnel is pitched to guide and drain the water at a constant, rapid flow, but it doesn't feel as though we are walking uphill. After the bore, the tunnel becomes square again, and smaller, but we can still easily walk upright. There is a second staircase with thirty steps. I have a feeling we are probably following the uphill slope of the golf course above. Michael and Peter with their walking sticks.

I'm curious to find the place where water is diverted from this drain into the remnant of the original Chedoke Creek, where I found the Arnold Palmer putter. Near the end of our journey, I think we've located it. A low dam spans the storm drain from one side to the other. The creek backs up behind the dam and flows over its top. A pipe is set into the tunnel wall on the reservoir side of the dam, its diameter line even with the top of the dam so that a constant, regulated portion of this water is diverted into the creek that runs through the neighbourhood. Simple, clever plumbing.

Light begins to brighten the way ahead. Are we there already? Time flies when you're splashing upstream in a cave. We step into a large, sunlit box of concrete. Steel grating that would make a penitentiary proud bars our ceiling and the wall ahead, which is two storeys high. We are standing in the reception room, the atrium of the Chedoke Storm Drain, at the base of the Niagara Escarpment, where the creek is captured and imprisoned after having fallen over the edge of the escarpment and travelled half a kilometre through a rock-strewn gorge.

The grate is piled high with the usual mix of urban and natural debris. Branches, rocks, a bicycle, plastic in many forms. We dawdle in the cage for a few minutes, then go with the flow back the way we came.

Our party of four hasn't yet had enough excitement for one day, so when we find ourselves standing in daylight again at the outfall end of the storm drain, we decide to wade upstream through the shallow, clear water of the open channel. Chedoke Expressway runs perhaps five metres above us. Traffic is only partially visible but very audible, yet the setting itself is so oddly magical in its battered-and-broken-concrete ambiance, overgrown with trees, bushes, weeds and wild flowers, that the sound soon recedes into the background.

A few metres from the first culvert tunnel entrance, the bed of the creek becomes stones and mud with small, red and pale grey pieces of shale. It's like walking on a real creek bottom rather than a concretized one. Where does all this earth come from? It must wash down from stretches where the creek runs free. The stony bed stays with us until we reach the other end of the culvert where the water deepens considerably, nearly to the top of our boots, and almost forces us to turn back. But not quite.

Walking through the channel beyond the culvert is as disorienting as being underground in a storm drain. With few visible points of reference, we can't locate ourselves in relation to the world above. Traffic flows at speed above us as we linger, lost along the way until a tall GO Transit commuter bus passes by on an exit ramp, the same bus that brought Nigel and Dorna to Hamilton from Toronto, and we know that we are beneath the Main Street West exit.

We may feel momentarily disoriented, but this is hardly citizen-uncharted territory. The sides of the second culvert are covered with spray-painted graffiti. At the upstream end of the culvert, the water pools again, deeper than the first. Forced to decide between taking soakers and turning back, we regretfully turn. I had hoped to encounter at least one of the five other creeks in the Chedoke watershed that join this channel, but will have to come back again later with waders.

Still not sated when we reach our starting point at the storm drain outfall, we opt to continue walking downstream. A double-culvert, one beside the other, takes us under the highway, and after a long straight run of the sloped, concrete channel, we come upon our first feeder stream. It pours over the lip of a two-metre-square culvert under the highway. This is the outfall end of the original Chedoke Creek, the garden feature creek, which begins in the bubbling pond where I found the Arnold Palmer putter. It flows parallel to the long storm drain that we four explored earlier, and is buried underground for most of that length. Here it rejoins its usurper and the five other creeks of the watershed.

Chedoke was diverted into the newly built storm drain in the early 1960s at the request of Westinghouse, and with some of their own money, as a way to deal with periodic flooding at their factory. The infrastructure project was planned to coincide with the construction of Chedoke Expressway. We can see through to the end of the culvert, to a small patch of daylight and greenery. I've stood at that other end, and wandered through the bit of nature that remains there in the form of a small flood plain wedged between the back of the former factory and the Main Street East off-ramp of the highway.

This paved-over and concretized meeting spot of waters in the valley is where Michael and Peter Hess, in that apocryphal story, stepped off the ancient portage highway and followed their own off-ramp up Chedoke Creek. Making this connection satisfies a desire I can only articulate by saying that, in my pact with this altered landscape, I take the creek's side.

The water in the channel is too shallow and fast-flowing to allow for fish, but the long, straight stretch we've been following now curves and enters a second double-culvert, where small fish begin to dart around our boots in numbers. The water slows and deepens and becomes too murky to see either fish or bottom, and too deep to walk any farther. We can't see the light at the end of it because the culvert curves. The channel to Cootes Paradise begins at the other end, but we've come as far as we can. The only way to finish the job is by canoe.

We call it a day. Three and a half hours. Tired. Exhilarated. We've had enough, but not nearly. Perfect, clear weather. Perfect, clear creek – except for this last stretch. Water legs. You don't realize how walking in water, with and against the flow, slows you down, inhibits and then inhabits your stride until you're contending only with gravity again.

Everyone hears about the crazy thing we did over the weekend because I can't keep my mouth shut.

Paul says, "Wow."

Dirk stares in puzzled disbelief and says, "That's your idea of fun?"

Peter tells me that he walked the storm drain when he was ten years old and it was being built. He did it twice: walked until it got too dark and he freaked out.

Lil remembers that she rode the channels and culverts with a friend in high school on a Honda 50cc motorcycle.

Once I start asking around, it becomes obvious that generations of kids have used, and continue to use, the channels and culverts and storm drain as a kind of playground.

"We go there when we're bored," Madeleine, who is fifteen, says.

Where have I been all these years? Not born here. Not raised in the neighbourhood. A Mountain boy.

Edward Berkelaar, a professor of environmental science at Redeemer University in Ancaster, and his student Janelle Vander Hout are collecting water samples this summer from the six creeks in the Chedoke watershed. The university has done similar sampling over the past few years, but instead of the usual single readings, they are taking readings every two weeks over a period of fourteen weeks. It's a summer project for Janelle.

The six waterfalls occur along a two-kilometre stretch of the escarpment that we access by starting where the Chedoke Radial Trail reaches the top of the escarpment. The Radial Trail is a popular walking path that climbs the escarpment face at an easy grade, using the abandoned right-of-way of the interurban streetcar system from the early twentieth century. The trail makes getting to the creeks and their waterfalls much easier than it otherwise would be. The railway crosses each of the creeks, channelling their water through pipes or tunnels under its bed, which in places is built five or more metres high.

Our first stop is Scenic Falls, west of the parking lot. There, the creek empties out of a square concrete culvert under the railway embankment and flows through an open channel of rock two metres high for one hundred metres before turning to fall over the edge of the escarpment. Edward and Janelle climb down into the rock channel to gather water samples and do testing. They measure for pH, dissolved oxygen and total dissolved solids at the creek. Back in the lab, Janelle will measure for nitrates, phosphorus, chloride, biological oxygen demand, E. coli and total coliform in the water samples. Taken together, they give an idea of the water quality. Too much of any of these will inhibit plant and animal life in the water. Their sources are mostly human. In the urban areas these creeks flow through, this includes road salts, lawn fertilizer, swimming pool runoff, laundry and dish detergents – anything that can make its way into the curbside storm drains. Then there are the illegal hookups that divert sewer system content into the storm drain system. Edward tells me that another student is spending the summer trying to develop a procedure to measure caffeine in water, which is the best indicator of human waste.

As a last act at each creek, its flow rate is recorded. Using a yardstick and a stopwatch, the scientists select a stretch of free-flowing water and measure its depth and width. Next, they choose a leaf from a bush or plant, place it in the creek and time how long it takes the leaf to float merrily, merrily down the stream. A leaf. Biology meets the inner child. I should have brought a clothespin.

After collecting the first samples, they carry the plastic bags of water the short distance back to the car. They will not make the same mistake as earlier, when they collected samples from each creek as they went and returned with backpacks fully loaded down with water. This time we pass by the creeks for Princess Falls, Mountview Falls, Sanatorium Falls, Westcliffe Falls and Cliffview Falls on the way to our second sample station, which happens to be Chedoke.

As we walk, Edward mentions that he told his thirteen- and ten-year-old sons the story of my storm drain adventure. His sons were enthralled. "Can we do that too?" He's not so enthusiastic, seems leery. This is not his idea of fun, for his sons or himself. Am I too blindly keen to realize others may consider walking underground through a pitch-dark concrete tunnel that is running with water a dubious, undesirable or even dangerous activity, and perhaps not strictly legal?

Chedoke Falls is half a kilometre up the gorge. Lower Chedoke Falls, two hundred metres ahead and five metres high, presents an immediate barrier to getting there, so Edward and Janelle take samples, test and measure the flow rate of the creek just above the grate at the intake of the storm drain – on the free side of the penitentiary bars our party of four stood inside of.

To get to each of the other falls, we have to clamber up a tricky terrain, over rocks of all sizes, often wet, slick rock, pitched to all angles, through bush, clinging to trees and tree limbs, sliding down mudbanks. It's a workout. In every case, we are rewarded with the minor magnificence of water cascading, terracing or falling in a ribbon over the lip of the escarpment above. Over the lip of a concrete storm drain, to be more accurate. All the creeks are captured under the neighbourhoods above. Yet somehow water has a way of rising above its engineered imprisonment and making it beautiful.

Between each set of falls, our feet and legs are treated to the gentle grade of the Chedoke Radial Trail. By the end of the morning, we are back at the parking lot.

A neighbour, Rick, rode his bicycle to school when he was a teenager, but since it was very uncool at that time to ride a bike to school, he parked it down in Chedoke Valley, just inside the mouth of the storm drain. So no one could see it.

One day, while he was in school, it rained hard. When he went to fetch it after school his bike was gone. The water levels had already returned to normal by then, so he walked downstream for a few hundred yards (they were still yards then, not metres) until he finally came upon the bike, wheels twisted into an unrideable shape, spokes stuffed with branches, plastic and other debris, mud drying on the entire mess.

He didn't have to worry about how uncool he was anymore.

On a biweekly schedule, Daniel and I take a hike together, or try to. This week we are set to explore the Chedoke Creek channel beside the 403 highway, upstream beyond the second deep pool that stopped the earlier party of four adventurers. I want to see where the other creeks in the watershed join the family in the valley. Instead of boots, we are wearing shorts and water sandals.

We enter off the same parking lot, and it feels magical all over again once we are standing in the channel's concrete waterway below the level of traffic. Nature is certainly having its way with the aging infrastructure, creating a weathered environment of imperfect, broken concrete, sending flora to burst forth from all the cracks. Some authority will want to reassert its control soon. We wade through the pool at the end of the first culvert and the deeper pool of the second, where my earlier party of four had turned back, and not much farther upstream, encounter our first feeder creek. My time with Edward and Janelle tells me that this must be the creek from Westcliffe Falls and Cliffview Falls, whose waters join not long after they fall. Those waters are buried underground for a few hundred metres, but then are freed to run through Chedoke Civic Golf Course. After its brief freedom through the golf course, the creek passes under a railway yard before emerging into a two-hundred-metre-long ravine that is tucked between the railway and the highway.

Daniel and I bow our heads and enter the culvert, wading and waddling into the ravine from its highway end, where we find ourselves standing in the doorway, between two cliffs carved through layered, red rock. It's a remote, unexpected and wild bit of nature in there, derelict yet deeply appealing; a debris-strewn, isolated remnant of a larger landscape, and sometime campsite locale for urban exiles and other homeless folk.

We enjoy exploring, and these biweekly walks are our personal, short-chapters version of El Camino de Santiago de Compostela, the ancient and popular pilgrimage route in southern Europe. So we claim. It is part of our pact with this landscape. We are each attracted to dimensions and purposes beyond the purely recreational in setting one foot in front of the other on our home terrain. Daniel has, coincidentally, just returned from a five-week stay in Spain, during which time he visited the cathedral where St. James is said to be buried in Santiago de Compostela, the destination of the pilgrimage. He was appalled, he says, by the ostentation of the cathedral itself and the high-end souvenir shops for pilgrims. It put him off the place entirely. He and Wendy didn't stay long. One saving grace was a man they met, in his seventies, who had walked the entire eight hundred kilometres five times and was preparing for his sixth. Decidedly not a spiritual tourist.

We are headed downstream again through the culvert as we talk. Bowing our way back to join the creek in the culvert. And bow we ought. To the current, to this hidden flow that pulls us along, that we are privy to.

Tall grasses pushing through the cracked concrete brush our arms in the narrowing channel. Another culvert appears ahead, its opening barred by a grate. The grate has been tampered with so it is not difficult to lift it high enough to pass under. We notice at once that the air inside the culvert is different, and dankly unpleasant. Water pours from a hole in the wall ahead of us. An outfall from the rail yards. Beyond it the creek seems stilled, no longer flowing. The bottom is soft underfoot as we move forward, almost mushy. The water feels gritty and a little oily.

We continue past the pouring hole, wade through the pooling water and finally reach an impassable grate at the tunnel's far end through which we can see, but not touch, green grass, trees and a narrow creek running freely over rocks. The land beyond the culvert. Is it possible that such a sylvan world should be feeding the one we are standing in? This is good news for a future morning pilgrimage. We'll later find this place on the map and continue our trek upstream through the neighbourhoods on the other side of this grate. For now we turn and head back, buoyed by the prospect, though the effect of this dismal cavern on our spirits does not lift until we are wading through the tall grasses again.

This is why our morning walks are pilgrimages. We willingly and purposely trek through places and landscapes that demoralize and test the spirit, that question what we as citizen-sapiens are up to, not least with these living threads of our landscape.

The map is rich with flags that indicate the different locations where Janelle gathered and tested samples from all the creeks of the three watersheds that drain into Cootes Paradise. She has collected samples from more than forty sites, and is making a presentation of her findings to a classroom of fellow students and professors at Redeemer University.

The flags are coloured green, orange or red, like traffic lights. Green means the creek has tested within all the guidelines and limits; orange indicates that is has exceeded more than one guideline or limit, or has badly exceeded one; and red says that it has exceeded all or all but one guideline or limit. The flags on the map before us are all green or orange. It's heartening. The creeks from the three watersheds are relatively clean.

All except for the creeks in the Chedoke watershed. Their seven flags are red, red, red, red, red, red and red. The clear-looking water we've been walking and wading through is deceptive. The E. coli count is particularly high. This is not entirely news. I have avoided mentioning that Chedoke Creek is sometimes referred to, in some quarters, for good reason, as Shit Creek. It's not the creek's fault.

The student who was trying to develop a procedure to measure caffeine in water also presents his findings, which are negative. Try as he might, he couldn't get a measure.

After the presentation, I want to hose down and sanitize my creek-walking bare legs and feet. Disinfect. Detoxify. Instead, Janelle and I make a date to hike the storm drain together. She is not put off by the facts or leery in the least. She's keen.

Piped under suburban streets and houses above the escarpment, and a golf course below. Channelled in a concrete ditch beside an expressway, its course determined by engineering rather than nature, above ground, as well as beneath it. One single, open and not-engineered section through a gorge after the falls that is impossible to experience unless you are inside the gorge, which is hard to get to.

Why pursue or follow this dirty little creek, this insignificant trickle? Witness and victim to our often sorrowful ways, it melds up and down in our geographically divided town, winds a connective thread between the upper and lower city, sings the six strings of its watershed, like a guitar gently weeping. Okay, maybe that's pushing it. But if the land is *alive and thinking*, then listening to and engaging with these living streams could help us retune our relationship with this place. By the song of its strings, we may be healed.

If only we could see those strings.

Daylighting is the term used when buried creeks are freed to run in the open again. A different relationship between running water and the urban environment can then be entered, one in which the landscape is part of the conversation. A number of cities around the world have done this with some of their creeks and rivers, and gotten surprising civic benefits from having water run beside their streets and past their front doors. And not just for kids with wooden clothespins.

I don't hold out much hope that Chedoke Creek will be cleaned and uncovered any day soon. Too many roads would have to be dug up, too many people would have to give up their properties or have them expropriated. Our way of thinking about how we urbanize a landscape would need to become more flexible, but our way is one-way rather than two-way, and often seems set in concrete. Until the next ice age arrives, not caring about our lines and nets, and pushes aside or uproots our constructions; scrubs down, restores and reshapes the geography; and perhaps reroutes the creeks entirely, this meandering line of words may be the only form of daylighting Chedoke sees.

I damn well better do its brave waters some justice.

The River that Flows Both Ways

Miss Vanden Top pulls down the map like a window blind in front of the blackboard for our grade three geography lesson. With her stick, she points out nine provinces and two territories, naming each as she goes. They are different colours, though some share the same pink or yellow or green. Many of the provinces are large, with straight or gently curved borderlines and some crooked sections. The ones on the far right are small and tightly bunched. Their lines are crooked because their borders are ocean. She points to and names the capital city of each province. These are represented by stars. Stars set inside a circle. All of us in the classroom live in a wheat-coloured province on the left side of the map. Its lines are straight, except for the bottom corner, which is angled and jagged like a mountain range. Our star is in the very middle of the province, and we live on it.

The stars float against the backdrop of the blackboard's outer space, in a constellation called *Canadensis.*

The world is handmade. Four hands altogether, belonging to a pair of twins, a girl and a boy, one dark, one light, one left-handed, the other right.

The right-handed twin smooths out the muddy lump of earth and makes it flat and uncluttered like a tabletop so that the view is unobstructed for miles in any direction. The other twin comes along and digs valleys, makes hills, builds the rocks up into great piles so you can't see very far at all and travel is a challenge.

One twin makes trees that are straight, with branches to match, and plants them in rows so no one gets lost in the woods. The other twin forks their branches and shifts the trees around and now the paths run crooked and it's easy to get confused, but also a little more interesting.

The first twin digs trenches and fills them with water and has the water flow both ways, handily allowing for travel in either direction. The second twin carves turns and meanderings into the creek and riverbeds and makes them run one way, then throws in rocks and waterfalls.

***Canadensis* is the name I am giving to that constellation now, on this return** trip, half a century later, to my grade three desk in Edmonton, Alberta. I imagine Miss Vanden Top poking a hole with a compass tip through each of the eleven stars, then holding the map against the light of the classroom windows to show us this relatively new constellation. That would be difficult. A student might have to help by holding one end of the map for her.

My hand shoots up, as do the hands of several other students.

The constellation is made of twin halves. Twins who do not look like each other at all. The left side is an upside-down dipper. The right side a jagged line.

Twins handcrafting the world is a story told (and retold) by the writer Thomas King in his book *The Truth About Stories*. It is the creation story he grew up with.

"The truth about stories is that that's all we are," he says.[1]

That creation story is a Haudenosaunee one. It is indigenous to this neck of the woods, the same woods Chedoke Creek winds its way through.

The spring that from the green root rises.

In my cosmology, "everything that lives is holy" – to borrow a line from the English poet William Blake. Creeks are living, therefore are holy. To take another futile stab at water description, a creek is water that loves what it does. It is a moving, staying-in-one-place, wet line that rolls over stone, twists around rock, hip-shifts to the contours of the landscape, lolls in low spots and falls from rocky heights in beauteous abandon.

A creek is a limber, lithe, winding country road to the river's highway. It is the journey and its destination.

A moving, staying-in-one-place, liquid fingertip that traces a path over the body of the earth. The feeling lingers on the body, from source to mouth, as long as the sun rises and the water flows.

Holy.

The creek that from the green spring rises rides through concrete veins.

I have sent a query to Hamilton City Hall, asking for the appropriate department, hoping to find someone who can give me directions. I want to know what route Chedoke Creek follows in its journey under the neighbourhoods on Hamilton Mountain.

Does the creek still recognize its way? Does it follow a path that resembles its natural twists and turns, or were trenches dug and an entirely new underground watercourse created when the housing developments were built?

From my grade three desk, a number of lessons that escaped me at the age of nine or ten are now coming home. For one, I see that each star at the end of Miss Vanden Top's pointer is located beside water. Running water or a body of water. Three of the stars dip their tips into the Atlantic Ocean, one the Pacific. Five are pinned to rivers: one each beside the Saints John and Lawrence, one at the confluence of the Red and the Assiniboine, another by Wascana Creek and one the Yukon River. Our star is hitched to the North Saskatchewan River; two other stars to lakes large enough to be called inland seas: Great Slave and Ontario. The major star in our national constellation is hung on the Ottawa River.

Had Miss Vanden Top pointed out these facts about cities and water? Did she tell us that the same is true for almost every town and village in the world? Did I forget, or was I not paying attention? Maybe I was distracted.

She gives the wooden bottom rail of the map a light, downward tug and holds on as the canvas rolls up into its curled slumber above the blackboard. She does this without haste, out of respect for the map, and for its temperamental mechanics. By the time it is fully wound, she is reaching high with one arm, her one heel not quite touching the floor.

The second lesson did not really escape me at the age of nine, but it is a bit of a surprise to look up now and realize that I am still in love with Miss Vanden Top.

DAYLIGHTING CHEDOKE

***Lost Rivers* is a documentary film about the rivers that flow under five** cities: Seoul; London; Brescia, in northern Italy; Yonkers, in New York State; and Toronto. Rivers and creeks have a long history of being buried in pipes, and then built over, as their cities have grown.

In Seoul a creek buried under roadways and an elevated highway for fifty years was daylighted. It is now a welcome urban feature both natural and contrived, since the creek has a low and irregular flow and water needs to be pumped into it constantly from other sources to maintain the illusion.

London has many creeks and rivers running under its many streets and buildings, and these draw enthusiasts who track the waters, draw maps and lead tours. Some waterways have been buried for centuries. They roar from under their manhole covers. This is also true in Brescia, where enthusiasts who call themselves *drainers* entered, explored and mapped the underground rivers when it was still illegal to do so. Their enthusiasm was catching, and the law was changed, and now public tours are given of the captured rivers and buried bridges and stone walls, some of which go back to Roman times.

Yonkers, just north of Manhattan in New York, daylighted a downtown portion of the Saw Mill River on which their former industrial economy had been based. The parking lot built over the river was removed. The project was controversial, but considered a success.

The Saw Mill flows into the Hudson River. The movie shows two men paddling a canoe under a highway overpass where the rivers meet as one of them sings, "I could be happy just spending my days on the river that flows both ways."[2] The Hudson, it turns out, is called the river that flows both ways because ocean tides travel upriver past Albany, over two hundred kilometres, twice a day, effectively reversing its flow.

The people who made *Lost Rivers* must have come from Toronto because the creek they champion in the film, and tried but failed to daylight, is insignificant compared to the waterways in these other cities (except perhaps Seoul). It's a lot like Chedoke Creek, actually.

At eight kilometres, Garrison Creek is about the same length as Chedoke. It travels south through a landscape crafted by the same glacial ages, from just above St. Clair Avenue through downtown Toronto, entering Lake Ontario beside Fort York, which gave the creek its name. It has two sources. One of these lies in a neighbourhood northwest of the corner of Bathurst Street and St. Clair Avenue. Of the two places where we lived when Mary and I were first married, one was at the bottom of that creek's ravine (a land feature our young legs failed to notice at the time, though it is fairly pronounced) on a street called Valewood. The second was a few blocks from the first, and only one half-block away from the creek's source. Holy.

"The creeks are calling to you," she says.

This is my desire, my backwards urge: to follow Chedoke Creek to its source. To walk upstream with Michael Hess and his ten-year-old son Peter. To accompany them, two hundred and thirty years later, even if it never happened, if it's not historically true they made the trip.

They wanted to plant themselves beside water, where water sprang from earth. They did plant, and the family grew, and branched forth, and branched farther, and bore much fruit. In the scrapbooks and photo albums that Marlene Gallagher showed me, there is frequent mention of a tree that grew beside the spring. A large tree. There is always a large tree. A tree that is both literal and figurative.

When I get there, I'd like to climb that tree.

Grindstone Creek, unlike Chedoke, is not in the Cootes Paradise watershed. Its mouth lies between the marsh and the bay, which is called Hamilton Harbour, so it is in the terroir of Chedoke and the other creeks that feed the marsh. Close enough.

It is also the opposite twin of Chedoke in that from source to mouth it flows (mostly) free.

Grindstone is a creek made for walking. The Royal Botanical Gardens has created a path that follows it upstream for more than a kilometre from its mouth. Beyond RBG property, the path goes through public park, then residential and undeveloped land until it reaches the waterfalls at the escarpment in Waterdown, a few kilometres away.

The residential area is called Hidden Valley. The valley is wide enough only for a two-lane road, a line of houses with shallow yards either side and the creek tucked against its steep west slope. The path picks up again where the road ends. Not far along the path, the creek bends and the water flows over a series of rocks.

Walking into this land- and soundscape for the first time, I stumbled onto another entry for a new lexicon for creeks. *Holy* was the first. The second is *prayer*. Not church or dinner table prayer, or the personal variety, but utterance, earth-speak, the run-on sentence that is sung between our world and one that is not visible. The water was saying something there, doing something: *being* on our behalf. Correction. Not *our* behalf, but in the name of life itself.

This is what I think: a creek is not only a living thing, but life itself.

City Hall has sent a map. The map came attached to an email from Udo Ehrenberg, a manager in the city's Public Works Department, Infrastructure Planning and Systems Design. My request concerning the present underground route of Chedoke Creek travelled through six people before it reached the right inbox. I'm impressed by the effort.

Poring over this map, my initial thought is that the first twin must have gotten a job with the city. This is the work of an orderly spirit. The map shows the creek as a yellow line that obediently follows streets and turns corners, with one-way arrows that indicate direction of flow. But the second twin must have had a hand in it too because, though it runs street-straight and all its meanderings are ninety degrees, the creek stays true to its original, winding path. It follows the lay of the land and drains the same watershed, the same catchment area, as it did before we got here.

Today's lesson is that a person could indeed follow Michael and Peter Hess's path upstream on the Mountain by walking not beside or through the water, but above it, on the streets and roads under which it now flows.

The creeks have been calling longer than I realize. I see by this map Udo sent that the neighbourhood where my parents purchased a house when we moved to Hamilton lies within the Chedoke Creek watershed.

What I saw when we arrived at our new home was a landscape being overtaken by housing. We were the front line in the suburban march through the former farm fields in which the streets were named after places in Florida. As if we really wanted to be there, not here. I did not think of our neighbourhood as a natural landscape. Nature remained in the form of a park at the end of our housing survey; undeveloped land with a low rock formation running through it. A sign eventually went up that named it Captain Cornelius Park. A row of mature black walnut trees crossed four of the properties in our six-house court, and we were told that these trees had lined the driveway of a farm. Before long two of the five were taken down. The neighbours in whose yards they stood didn't want them there. It was clear that in this landscape of streets and houses and yards that had been created for us we could make and unmake our portions of it as we saw fit. As the housing development continued to grow around and beyond us, the grade eight take-home lesson for me was that a made landscape obliterates and replaces the natural one that precedes it. It's one thing to move into an established, leafy suburban neighbourhood, but when you are twelve years old and are witness to its cutting edge, it can mean serious grief, especially if you were able to play in and explore the fields and woods before the first wave of bulldozers arrived. You may find yourself waking up in the morning sad and miserable. With a feeling of dread. For no apparent reason. You have everything you need: a roof over your head, a warm bed, food on the table, a family that loves you, friends.

Maybe it would have helped had I known that the slope in our backyard was there not solely to frustrate my father by sending the neighbour's yard's runoff into his row of young poplars, but that it tipped the water toward the street catchments of a living creek that called and claimed as its own our lawns and driveways, as well as our roofs, and our courts, lanes, places and drives. Maybe I would not have felt so displaced, so *un*placed, if someone like Miss Vanden Top could have pulled down, in front of the blackboard, a map like the one Udo sent me and explained that we dwellers of housing developments and cities still have a connection to and with a natural landscape; that an urban countryside still runs with water.

Someone should have told me that we remain in song, though the singing be so low it murmurs only through a grate.

The world was made in the telling. The story that I heard while growing up did not include twins, although there were brothers, Cain and Abel, who came along a little later and didn't get on with each other any better than the two in Thomas King's retelling.

This world-making is voice-activated. The voice speaks, *Let's have some light; Let's have some trees*, and things begin to happen. What didn't exist, does now. It's all in the telling. Sit around the campfire on a long weekend beside the lake and listen as the sky slowly revolves over your head, words are spoken and the world is made.

The truth about stories is that that's all we are.

The creating of the world is enough work for one week and results in a garden, a paradise where the animals are your friends. Humans are a bit of an afterthought. First one, then another. They are well matched but don't always agree, and the upshot of the drama that unfolds is the end of garden living, though we knew all along it was too good to last. That much was written in the stars.

Whoever invented this story must have lived in a city. They must have lived in close quarters with enough other people long enough to have learned first-hand what occurs when half the population doesn't care what happens downstream, and the other half has children or grandchildren who live there.

The first one to tell this story must have lived in the city but spent summers as a child in a canoe on a lake or northern river.

Miss Vanden Top became my lost love one day in grade three when I and two of my classmates were standing in the doorway to our classroom waiting for her return from the Teacher's Lounge. Just as she reached the door, extending one arm to gently corral us inside, I fainted. For no apparent reason. She caught me in her arms before I hit the floor, I was later told, and carried me across the hallway, where she handed my limp form to a male teacher who carried me to the nurse's office.

What? She couldn't carry me there herself? I learned then that she and the male teacher were to be married at the end of the school year. She had already chosen. All others would be handed off to her selected for him to deal with.

She could no longer be sister or Joan of Arc. She could not be my guide through our northern wilderness. It was time to launch out on my own.

Hamilton was not a star on that grade-school map. One of the lesser lights in the constellation *Canadensis*, its brightness was further dimmed by some pretty serious industrial pollution.

In its early settlement period as Head-of-the-Lake, the first Euro-migrants here cohabited and intermarried with First Nations people who still lived, worked and freely wandered land and water, and with the wolves and cougars and bears, and the rattlesnakes that lived in caves in the escarpment, and the salmon that filled the bay.

Early days always sound like paradise.

Leather tanning, the making of coal oil, overfishing, a steel industry, sewage. The litany of troubles our waters have suffered over the past two centuries is long. There is also a matching list of early and ongoing efforts made in response. But that's just it:

the efforts are usually reactive rather than proactive and relational. John Kerr, the first fishery inspector, hired in the 1860s, crouched on the edge of the escarpment with a pair of binoculars, observing and nabbing poachers on the bayshore. Fish stocks were already diminishing noticeably. Beginning in the 1990s, a series of CSOs (Combined Sewer Overflow tanks) were constructed to help prevent raw sewage from entering Hamilton Harbour or Cootes Paradise when the storm drain system is overwhelmed during one of those massive storms that have lately become more frequent.

The sad fact is that there is no return to paradise. And yet, though I'd prefer a clean and free-flowing Chedoke Creek to run through the suburban Mountain neighbourhoods, I find the map that Udo sent a thing of beauty. A hidden beauty. It may be the product of expediency rather than love, but it follows the lay of the land and signifies a taking-care, a tending of this garden.

The biologist Edward Berkelaar, who, with his student Janelle Vander Hout, led the water sampling at all the escarpment waterfalls of the Chedoke watershed, wants to make the results public. He would like to do so in a way that allows the information to reach a wide audience without the science causing eyes to glaze over. So together with another Redeemer University biologist, Darren Brouwer, we are meeting with Tys Theysmeyer in his capacity as Head of Natural Lands at the Royal Botanical Gardens, which owns and manages much of the land around Cootes Paradise. The RBG has already spent years finding ways to improve water life in its holdings. Tys has brought a friend, Alan Hansell, from the Stewards of Cootes Watershed, a group that pulls on hip waders and hauls garbage out of the streams. More than one hundred tons has been pulled out so far, much of which is "historic," as Alan calls it. Junk that has been there for a generation or more.

We meet at 7:30 in the morning at a downtown café, and put together an evening program of story, science and service which weaves together our various projects. It will take place two months from now. We leave the café, each with our allotted time slot and a list of responsibilities toward making it all happen.

This watershed thing is still rather new to me, but as we sit together talking, I slowly begin to realize that the café we're sitting in, which is below the escarpment and near the city's downtown, in my old neighbourhood, in fact, also lies within the Chedoke watershed. For the quarter century that we lived in this neighbourhood, my wife and I were living in the watershed. And even now, in our new, old brick house on the Iroquois Bar, the glacial landform that runs crosswise through the downtown, we are perched at a point where the water flows two ways: beyond our backyard it runs toward the bay; from our front it goes toward Chedoke Creek. Holy.

These creeks have been playing me all along. Playing me like a song.

Our clothespin boats were swept into the curbside storm drain and carried who-knows-where by the big creek, North Saskatchewan. The summer my parents packed us up and we left Edmonton, I went down to the river to launch another gift.

During Canada's centennial year, 1967, the government encouraged everyone, young and old, to invent their own projects to complement the many federal, provincial and civic projects that were on the go. The stars were aligning for a country one hundred years young. Disparate parts and pieces of the Dominion would come to see themselves as one. And, for the time, they did.

The encouragement began a year ahead, in 1966. For my project, I outfitted a foot-long, red plastic canoe with a waterproof package tied securely into place. Inside the package was my name and address (. . . *Edmonton, Alberta, Canada, North America, Earth, Solar System, Milky Way, Universe*), a plea to make contact if found and a few valuable trinkets as incentive. I was in thrall to *Paddle to the Sea*.[3]

Working my way down the steep slope to river's edge, I carefully placed the canoe in the water. It lingered, turned sideways, backwards, then caught the current and floated downriver.

Lynbrook Drive and Millbank Place are streets in the heart of the Hamilton Mountain suburbs that were built in the 1960s, part of the same housing wave whose crest our family rode when we arrived. I drove up there this morning. My *Hamilton Spectator* paper route took in those two streets, as well as others in the neighbourhood. The delivery truck dropped the paper bundles at the bus stop on Mohawk Road at Millbank, across the street from the nineteenth-century, brick Mohawk Trail Schoolhouse (which was recently moved and relocated). I delivered to sixty-six houses, although new addresses were added to the route on what seemed like a daily basis as people moved into their just completed houses south of Mohawk.

Lynbrook Drive ran parallel to Mohawk and ended one house west of Millbank, which connected the two. Lynbrook dead-ended where two horizontal planks nailed between two wooden posts, painted yellow, barred the way to the field beyond. I drove there because while comparing the route of a Chedoke Creek feeder stream on the 1875 map with a current city map, the yellow barrier had rematerialized out of the November mist of my memory. The older map showed the feeder crossing Mohawk Road beside the school, having flowed down Millbank after turning left from Lynbrook. I remembered standing, looking out over the field after delivering to the one lone house, and, as I stood, the yellow barrier slowly emerged before me from the cranial fog, and with it a creek running through the field like an extension of the road itself, flowing toward me and entering a storm drain under my feet. Surprise, surprise.

This is just the kind of memory I would expect myself to invent, so I asked a higher authority. My brother helped with the paper route and later took it over, and is now a historian who remembers, more clearly than his elder siblings, family events that took place when he was two years old. He confirms the creek.

A creek long buried underground and in personal memory is daylighted.

Is the creek the reason these street names are so waterlogged, why they are Lyn*brook*, and Mill*bank*?

I drove through the yellow barrier that exists in memory only, and followed Lynbrook Drive as it curved like the winding creek that now flows captured in a storm drain beneath it.

The creek was not the only reason for my visit to the neighbourhood. The Smith family, who accompanied the Hess, Kribs and Rymal families from Pennsylvania to Head-of-the-Lake, also looked for a steady supply of water and chose this parcel of land on which to rebuild their lives. When their friend Michael Hess died in 1804, the Smiths donated a small parcel of their land for his burial on a low hill overlooking the stream now under Lynbrook. Then, when Michael's wife, Charity, died on the same day four years later, she was laid beside him. Hamilton Mountain had its first cemetery.

The cemetery still exists, but is hidden. A narrow, grass lane leads between the backyards of houses to a small knoll, roughly thirty metres square, behind a church at the southeast corner of Mohawk Road and Garth Street. Once before, in haste and bad weather, I visited and found no headstones or any clue that this was a cemetery, though the very fact that a small plot of land, not infringed upon, existed in the middle of a parcelled landscape defined by fencing, suggested something untouchable, if not sacred. The only other such parcels of defined yet untouched urban land that I know of are contaminated sites, which identify themselves with their telltale groundwater-monitoring well pipes: ten-centimetre-square metal posts with a lock on top, standing about the same height as a headstone.

A patch of bush grows at the top of the small hill beside which stands a tall, old bur oak, one metre in diameter, in ill health. Shagbark hickory trees grow two or three together here and there, at a distance, respectfully, allowing the oak to preside. This time out, at leisure and in nicer weather, I stuck my head inside the patch of bush and, thinking of ticks, saw what the oak was presiding over or, perhaps, protecting: the broken tooth of a headstone protruding eight inches out of the ground. Then beside this broken tooth, a second, and also a third, in diminishing heights.

There were no weather-worn inscriptions to be seen on these jagged fragments, but I claimed the first for Michael, the second for Charity and the third stone for whatever it takes to keep the waters freely flowing between our past and current lives in this landscape.

This is how the river flows both ways.

It rises and springs out of the earth in defiance of gravity, then runs as swiftly as it is able, in accord with the humps and tiltings of the earth, chasing after a kiss.

With its mouth, it plants a big wet one on the body of water from which our two travellers set out on their adventure, journeying upstream, seeking the source.

Our travellers seek to know where all this love comes from.

The plastic red canoe of my childhood finds its present incarnation in the form of Daniel's deep-red Kelvar Beach Marine (Made-in-Hamilton) vessel, which we carry from the car and set into the water of Cootes Paradise at Princess Point.

Thus begins the official first leg in the journey from creek mouth to Mountain spring. From this spot, Michael Hess and his ten-year-old son Peter launch their journey to find the stream's source atop the escarpment, where they will build Spring Farm out of the ruins of their life in the United States. After travelling days and weeks over land and lake, the pair stop at last at this portage head at the end of the bay where another refugee, Richard Beasley, has set up his trading post. They gather their minimal gear together and mount the height of the Iroquois Bar, where, from its crest, they spy the creek flowing through an arm of the marsh below them.

"I like it here," says the young Peter.

I like it here myself, though much has changed. The marshy arm is Chedoke Valley entering Cootes Paradise. Once called Beasley's Hollow, it is now buried under a decommissioned city dump that is the clay-capped, grass-carpeted home to several soccer and baseball fields, and the six lanes of the Chedoke

Expressway. In this heavily engineered landscape, Chedoke Creek has become a straight, ten-metre-wide, one-kilometre-long channel beside the highway.

It's a beautiful, late summer morning. Cool, cloudless and bright. We paddle to the mouth of the creek through patches of lily pads and past conclaves of cormorants perched on the dead arms of fallen trees that have washed into the marsh, then glide under a bridge for the Waterfront Trail. The bridge acts as gateway to the creek. Almost immediately Daniel spots a young, black-crowned night heron, the first of several, hiding in the branches of one of the trees that overhang the canal. The heron flies off upstream as we approach. Ahead of us, floating cormorants keep their distance, now and again also taking flight farther upstream. We feel bird-led, or lured. We note a beaver lodge to our right, on the west bank. Yes, a beaver lodge. We note a red-and-blue nylon tent set up in the reeds on the same bank, with an armchair beside it, facing the sunrise.

The water is murky and almost viscous-looking. On our left, a steep embankment rises four metres to the highway. Alan of the Stewards of Cootes Watershed told a story at our café meeting of his group's cleanup of this creek channel and of finding sizable objects, such as bedsprings and appliances, exposed in the embankment like the bones of dinosaurs in a Badlands coulee. We keep our eyes peeled for discovery.

At regular intervals, metal drain pipes protrude from the embankment. The pipes come in two sizes: half a metre and a full metre in diameter. We paddle under the bridge that gives access to the playing fields of the former dump and smell diesel oil. It's brief but distinct, and strong. A little farther upstream, the creek becomes shallow and less murky. The bottom is visible, though it is covered in a brownish-green film. We inch around the crown of a fallen tree and enter a more open area, where the bed is wider and there is even a little island. This is less channel and more creek-like. The water is too shallow to paddle, so we disembark to pull the canoe forward. Less than one hundred metres ahead is the large, two-mouthed, concrete tunnel that carries the creek under the highway. The creek bed is fairly firm, but I step in a spot that draws my feet farther down into the muck and must struggle to lift one, then the other foot free. The old fear instantly returns. Men, women, children and animals in childhood movies and TV shows who slowly sink out of sight in quicksand, leaving only mud bubbles behind. My worst nightmare. "What if I get sucked down and disappear forever in this muck?" I ask Daniel. "I'll have to call your wife and tell her that we lost you," he says.

The highway and its access ramps roar and curve above us, two apartment buildings tower through the overhanging trees, and the slopes of concrete channel sides lie in broken slabs, while suspect water flows around our feet. Is it despite or because of these realities that we both find this sorry place so disarmingly lovely? Somehow, nature is achieving something here anyway.

The water deepens again. We step back into the canoe and choose the left half of the dual, wide-mouthed tunnel, entering the heart of linear thought and lines drawn on paper, of engineering and infrastructure. The first twin wins this one, hands down. The smooth concrete walls, four metres apart, are stained but in very good condition. The bottom, a metre below, is also concrete. The wall to our right soon opens to a series of three or four pillars, which provide a view of the identical tunnel next to us. The tunnel curves, and, as we follow, total darkness envelopes us. Anything could be lurking ahead, above or beside. At the same time, there is a pervasive sense of total lifelessness here, so theoretically there's nothing to be scared of but the dark. Says Daniel. Then something splashes in the water.

When we finally pull out the phone-flashlight, we can see that halfway up the left wall a wide feeder tunnel is joining ours. Beside it a ladder climbs to a grate through which some little light is falling. The water is too shallow to continue by canoe, but neither of us has the slightest inclination to get out and walk on. No matter. I know where we are. This curve in the tunnel. We've reached the point of contact with the trek Nigel, Dorna, Mary and I made coming from the other end. Mission accomplished.

We try to turn ourselves around, but the canoe is longer than the tunnel is wide. After a comic routine in which our vessel gets wedged sideways between the walls, we paddle in reverse and back ourselves downstream.

The smell of diesel recurs downstream, strongest where a galvanized metal pipe about a metre in diameter emerges from the highway embankment. A small stream flows over its lip into the creek, about two gallons a minute. The pipe bottom is crusted with a thick overlay of a tan-coloured gunk. Combined with the diesel smell, it's pretty unsavoury. Loathsome, even.

Has Chedoke Creek not borne enough grief or carried sorrows enough? I will call someone about it when I get home.

Six hours later, I am standing in the parking lot behind the French-language high school on Macklin Street, which runs parallel to the Chedoke Creek channel, on the west side of the creek. Chris Banitsiotis, from Hamilton Public Works (*Water is Life*), is already parked there in his Environmental Monitoring and Enforcement van. "To report Spills 24/7 Call 905-540-5188." This sounds straightforward enough, but try finding that number when you are looking for it. In the end, I sent an email to Udo Ehrenberg, the city manager who sent me the map, asking who to contact about a possible spill. That email followed the same swift and expeditious route as my previous one, going from one person to the next until it landed in Chris's inbox. I'm impressed all over again.

We meet in the parking lot because when Daniel and I smelled the pipe we used the brick wall of the school as our landmark. Chris wants me to show him where the pipe is, while at the same time indicating to me that the pipe does not likely belong to the city. If it comes from under the highway, then it is a drainage outfall belonging to the provincial Ministry of Transportation. The city has its own leachate drainage system under the former dump on the other side of the highway, and he shows me a map of this on his laptop computer. The only city storm drain emptying into the creek comes from this side, from the roofs and parking lots of the school, the retirement home next door and the city arena across the street.

The creek is invisible through the trees and bushes that line its shore, and a chain-link fence prevents anyone getting any closer. While Chris calls the Ministry of the Environment, I walk the length of the parking lot and eventually find a break in the foliage that miraculously provides a perfect view of the offending pipe.

He is talking with someone from the ministry when I return to the van. It seems clear that his days are spent navigating the terrain between various levels of provincial and municipal jurisdictions, industrial and residential environmental events, and people like me who call up on the phone. It takes time. Spills and other environmental events are not as straightforward to unravel as they may seem.

While still on the phone, he relays questions to me: Was there a rainbow-coloured sheen on the water around the outfall of the pipe? Did we notice a smell or sheen on the canoe or paddles when we returned to the car? No to both questions. Perhaps then we are dealing with a smell only, rather than a spill. This is a relief. But where is the smell coming from?

Phone calls earlier in the day had put me in touch with the district officer of the Ministry of the Environment. Chris learned from his phone call that Mike Durst, the officer, will be following up. He promises that I will be kept in the loop.

"The Story of the Chedoke Watershed," as our event was advertised, was gratifyingly well attended. Edward, Tys, Alan, myself and a late addition, Mark Bainbridge, who is a Director at Hamilton Water and Wastewater, each had our moment in the spotlight before a near-capacity audience in the auditorium of the *Hamilton Spectator* building, a brick bunker that overlooks Chedoke Valley, and made our individual pitches for the life of our local waterways.

A few weeks later, I joined Alan and the Stewards of Cootes Watershed for one of their Sunday morning cleanups. Perhaps thirty of us gathered in a parking lot where we were divided into two groups. Our team was sent to clear a runoff ravine that feeds into Chedoke Creek almost as far up-valley as Daniel and I had reached. Because a few private properties backed onto it, the ravine contained yard and household debris and a dismantled deck of pressure-treated wood, in addition to the usual car tires, scrap steel, glass bottles and plastic in every shape, colour and form, whole and in shards. Trash new and historic. The pile we created on the street was two metres high, two metres deep and ten metres long. Team members would later come with a pickup truck.

The work is rewarding and hopeless because you know you'll never get it all, and there's always more where that came from. You're picking up after others. At the same time, though, you're doing something, not passively accepting nature's fate as dumping ground. There are sections of the twenty-plus creeks in the watershed where, because of the obstructions now removed by the stewards, Tys has observed fish beginning to spawn again.

From the mouth of Chedoke Creek, after our excursion upstream, Daniel and I paddled farther into Cootes Paradise, around Princess Point and into the Westdale Inlet.

The three-hundred-metre-long inlet is as quiet with wind, water and trees as the Chedoke Creek channel is loud with highway traffic. Daniel's bird-eye spotted a great blue heron, a green heron, a spotted sandpiper, terns, an osprey and a falcon being chased by or chasing another, smaller bird. We spotted a huddle of turtles sunning on a log. One by one, they slipped into the water for cover as we cautiously approached.

Dragging our paddles and lolling about the lily pads. This must have been what life was like in the creek too, once upon a time. When it was Beasley's Hollow. The sheer abundance of life here is actually pretty recent. Lily pads can flourish because the fishway at the mouth of the Desjardins Canal catches the carp before they enter Cootes Paradise, where one by one, RBG staff return them to the bay. The notorious bottom-feeders had for years prevented marsh plants of any kind from growing. The water itself is also much cleaner, the result of efforts in the Cootes Paradise watershed as a whole and the Dundas Waste-water Treatment Plant upstream. "Paradise Regained" was a slogan the RBG used to move people and money into action. The work is far from over, but maybe one day the billboard map beside the canoe launch area at Princess Point will need repainting because of these ongoing efforts. The map shows hot spots in the marsh, the places where pollution is greatest. The area at the mouth of Chedoke Creek is coloured the same bright red as Daniel's Kevlar canoe.

Mike Durst from the Hamilton office of the provincial Ministry of the Environment called to say he had been to the site, had seen the pipe, which was almost impossible to get to except by canoe, and that he wasn't too worried.

Based on what I had told Chris, who had relayed to him that there was no rainbow sheen in the water and no lingering smell on the canoe or paddles, this was not a situation in which the water was being contaminated. The diesel smell could be diesel gas from highway construction nearby.

Smells happen all the time, he said.

He was off for a week of vacation, but would make a few phone calls when he returned to see if he could discover where the pipe was coming from and get back to me because, as I told him, I was still curious, and concerned. I was curious about the smell and about where the continuous flow of water originated. Concerned because smells may happen all the time, but that doesn't mean they are innocent.

In the days that followed our Chedoke channel paddle, I drove the King Street ramp onto Highway 403 several times, eyes resensitized to the lost and confusing landscape of highway, creek channel and decommissioned city dump.

Tracking down the source of the mysterious and smelly outfall pipe would mean exploring the area on the other side of the highway from the channel, so I parked beyond the French high school on Macklin Street and walked across the bridge that spans the channel, through the tunnel under the highway and up into Kay Drage Park. Kay Drage is the name given to the capped dump and its soccer and baseball fields.

The playing fields are higher than the highway by a number of metres. A steep slope plunges into a ditch that runs beside it. The slope is home to a dense wall of foliage. Sumac. Tree of heaven. Manitoba maple. The usual suspects. I was looking for the landmark corner of the French school's brick wall that would tell me where the outfall end of the pipe was, but couldn't see through.

Upstream and beyond the playing fields, a gap in the chain-link fence gave access into a surprisingly large area past the dump and between the highway ramp to the west, Catholic cathedral to the south and train tracks to the east. Two hectares of *terrain vague*.[4] A man approached, walking through the weeds with his dog, which he stopped to leash. We talked briefly. A frequent traveller of these paths, which are many, he said, he's regularly seen deer and coyote, and hawks flying overhead.

His description had me anticipating a nature walk, but nature was mostly weeds and bush, with desolate groupings of Siberian elm, their small leaves riddled with moth holes. A small stand of what I think were young butternut trees appeared sickly too. The area is lower than the dump but higher than the creek level, and since this was all once part of the marsh, it must also be landfill of some kind. At the bottom of a small glen created by the highway ramp and cathedral parking lot, a square blue pipe stood a metre high and ten centimetres square with a lock on the top. Groundwater monitoring well.

A deer path led through the woods and eventually joined a more well-trodden footpath closer to the railway tracks. In one direction, the path climbed to a gate in a chain-link fence that opened to the cathedral parking lot, in the other, it probed deeper into this terrain. Forking from the second onto another, narrower path, I soon found myself in someone's outdoor living room. A small, round hut, covered in standing reeds, stood on one side of a firepit; on the other was a wooden shed, two metres tall, with a locked door. Water jugs and cooking pots were set in scattered holes around the site. I took a few photographs, felt intrusive and moved on.

In one last attempt to see through the bush at the top of the highway embankment, I crawled through a hole in the fence and scrabbled through its dense maze for a clear vantage. I found a broken microwave oven and other urban jetsam, but made little headway, and saw no more than I had earlier. A vain attempt in a no man's land. No land at all. I climbed out and leaned my chin and elbows on the top bar of the chain-link fence, gazing blindly over the noisy highway, drowning in sound. It doesn't matter. I'm here for the creek. The creek knows. This is how things are now. Things were not always like this. They will not always be like this.

Sometimes you need to pause in the midst of it all and reacquaint yourself with the long view.

Chedoke Gorge encourages the long view. This half-kilometre-long gash in the face of the Niagara Escarpment, a narrow canyon, a rock-walled, boulder-strewn cul-de-sac with the creek running along its bottom and a cascade, Chedoke Falls, at its dead end, is the product of twelve to fifteen thousand years of postglacial erosion and the welcome efforts of the second, more mischievous twin.

It is also a feature as close to being in its natural state as you could hope to find in an urban setting that has seen two centuries of Euro-human habitation.

So sit and mull to heart's content. Listen to water give tongue by rock. Sing along. Make, or reaffirm, your pact.

Stories like this one, they step in and out of the current. The passage of two hundred years has freed them from the constraints of time and space, allowing them to stop and start, to travel upstream or down when the opportunity arises and the spirit moves. This is my rationale for pursuing/following Chedoke Creek in such random fashion, in a quest to piece together Michael and Peter Hess's journey from mouth to source.

Chedoke is approximately eight kilometres long, and divided into five distinct sections.

From mouth to source, these sections are: the channel beside the highway (three-quarters of a kilometre), *check*; the slope-sided concrete channels, also beside the highway (three-quarters of a kilometre), *check*; the storm drain beneath the Chedoke golf course (one and a half kilometres), *check*; the gorge (half a kilometre); and finally across the Mountain, underground in a storm drain, from the falls to the spring of Spring Farm (four kilometres).

The three downstream sections have been travelled, and only two sections more remain: the gorge and the creek's hidden Mountain run.

Best for last.

The Mountain leg of the journey begins at the concrete bridge rail that over-looks the top of Chedoke Falls, where Daniel and I lean to watch the creek emerge from the large drain pipe, drop onto a lip of rock three metres lower, then plunge thirty metres into a pool before enjoying its day-lit flow between rock walls. If there weren't so many trees blocking the view, Chedoke House would be visible half a kilometre downstream, perched on the edge where the gorge breaches the escarpment face.

The railing separates the road, Scenic Drive, from the drop. Sated with the view, we turn and cross the street into Colquhoun Park. We are chasing father and son through suburban streets today, yet immediately to the right stands another holdover from the nineteenth century, a stone house oriented with a view of both creek and falls from the days when the creek ran freely. Those days ended for this park in the 1960s, around the same time the storm drain under the golf course below was built. I know someone who lived here as a child, and who still lives beside the park, who launched boats and played in the open creek. If any portion of Chedoke were a candidate for daylighting, this open expanse of park and grass is it. For now Daniel and I make do by putting our ears to the perforated, cast-iron grate in the grass, "McCoy Foundry 1956," to hear the water flow. McCoy Foundry was the main supplier to the city of storm and sewer grates for a generation and more.

The city's storm drain map directs our feet beyond the park, south on West 23rd Street, then east after one block onto Bendamere Avenue, across Buchanan Park and along South Bend Road. We try to match the lay of the land to the original route of the creek and to the piped creek, but there's no simple, one-to-one correspondence. A swell rises where there ought to be a dip, a dip where the land should swell. The two occasionally come together, but the landscape does not consistently behave as though a creek were running through it.

Along the way, we take note of downspouts. Newly built houses in this town are required to route rainwater onto the garden or lawn. Because its storm drain system has gotten overwhelmed several times recently during severe weather events, the city has asked owners of earlier-built houses in suburbs like these to disconnect their downspouts from the system and send the water onto their gardens or lawns. Most of the houses still have their downspouts connected directly into the storm drain.

What's with these Mountain folk?

These Mountain folk have other plumbing issues that need attention as well.

The seven flags at Janelle's presentation on the creeks in the Chedoke watershed were coloured red because of E. coli in the water, and the E. coli is present largely because of what are called sanitary sewer lateral cross-connections, or illegal hookups, which send sewage into the storm drain system.

Some of the hookups are done on purpose for convenience, some in error. A cross-connection is found on average every five hundred metres of storm sewer – and these are corrected at no expense to the homeowner. The city is not interested in laying blame at this time. Almost two hundred have been corrected in the Chedoke catchment since the program began. It's not just Mountain folk, however. The problem is common across the city and to cities across North America. Boston has been working on correcting the problem for two or more decades and has discovered that it never goes away. New sanitary sewer lateral cross-connections keep popping up.

A certain percentage of the population always wants to pee into the stream.

We interrupt our pursuit so that I can show Daniel the cemetery with three broken-tooth tombstones: one for Michael Hess, one for Charity Hess and one to keep the creek flowing both ways, and uncontaminated, between past and present. I have since learned that it has a name: Hess Family Burial Ground.

A shadow lurks within the cedar bush in the northeast corner of the site. It is immediately visible this time, but somehow the one complete headstone still standing had avoided detection on my earlier visit. We part the branches and read its carved inscription: "In memory of Jane Wife of John Snider who departed this life on the 11th day of May 1820 Born 11th April 1781."

Is Jane somehow family too? She had just turned thirty-nine when she died, and was only nine years old when the first five families arrived from Pennsylvania. A contemporary of Michael Hess's son Peter. Was she among the group of fifty?

We chat briefly with an older Italian woman standing nearby in her yard, who offers nodding assurances to Daniel and me that the dead are doing an excellent job watching over her and the neighbours. Are there more headstones hiding somewhere? Daniel engages a second backyard neighbour in conversation, an immigrant Scot who says that he moved into the new housing development as it was being built in 1969, fresh off the plane. His house stands in a small court off Lynbrook Drive and its buried creek. The cemetery was full of headstones at the time, he tells us, many of them with Hess names on them. But at some point during all the construction, the church allowed trucks to use the cemetery as a thoroughfare, and the stones were removed and brought to the church basement, where they were to be stored, and supposedly are still. The church maintains the cemetery, but no one he has talked to knows anything about the missing headstones. There's more than two hundred people buried here, he says. It's sacred ground.

Interesting story. Curious and kind of terrible. But it's time for us to go. This was supposed to be a quick side trip. It proves difficult to disengage from the friendly Scot's web of words, however. As we wave, turn and start walking, he asks, "Do you want to hear a ghost story?" "Not now. Next time." We are leaving earshot when he adds, apologetically, with a smile and a shrug, "I was a high-school teacher: I talk."

I drive slowly from the parking lot onto Mohawk Road, careful to note the church name in order to make contact later and try to chase down those lost headstones. Church of the Resurrection.

Does a creek cast its own spell over its own watershed?

There is a distinct feeling to these suburban neighbourhoods that Chedoke Creek winds through, though it's hard to separate this feeling from the fact that almost everything was built in the 1950s and '60s, and has the same architectural stamp. And it's such familiar territory to me.

Yet I wonder. Does a creek cast a spell? Do the tendrils of its feeder streams, its fluvial root system, lay a grip on the landscape that can be felt in the air – even after urban development has captured and hidden it? Does it create its own atmosphere, a kind of weather, whether running in broad daylight or buried in the ground?

We left Michael and Peter waiting for the bus at the corner of South Bend Road and West 5th Street. The storm drain map now bids us walk south on West 5th, then turn left after one block onto Richwill Road and travel a few blocks east to Upper James Street. Again, the landscape is too subtle in its undulations to make any convincing correspondence between original creek bed and what we see around us. At Upper James, we turn south and walk upstream to the intersection at Mohawk Road: four lanes of traffic, plus right and left turning lanes, meeting four lanes of traffic, plus turning lanes. A piercing siren halts all movement as an ambulance races through the intersection in a wide arcing turn, after which the stop-start rhythm of streaming cars resumes.

I note for Daniel the Dominion grocery store on the north-west corner, with strip mall attached, where my mother shopped, though the store is under a different name now. It's the corner where our travellers from Pennsylvania camped in 1788.

Across the street stood the Hess Inn, owned and operated by Michael Hess's grandson Jacob in the mid-nineteenth century. One hundred years later, this became the site of a drive-in restaurant called the Millionaire, or a drive *inn*, as its landmark, ten-metre-tall sign advertised. The Millionaire stood in a cut-away jacket with tails, spats on his footwear, chef's hat on his head and a thin handlebar moustache, offering to the world a mug of root beer as large as his torso in one hand and an equally grand hamburger in the other, while standing on a pile of gold coins. He stood in florescent and neon greeting when our family first arrived, but has long since taken his coin elsewhere, leaving a casket-flat strip mall in his place.

Across Upper James stood the department store, now a Canadian Tire, where my three-speed bicycle was stolen when I was thirteen and buying a record album inside. No, I hadn't locked it. Who did? The middle of the Canadian Tire's prairie parking lot is where, as I told my small audience at the Terryberry Library last year, I first thought the source for Chedoke Creek was located. I was partly correct. The creek begins a kilometre upstream, but did travel under where the parking lot is now. The parking lot is designed to shed its water toward storm grates, which are connected to storm drains, which empty into the creek. In my imagination, a spring bubbled forth from one of the storm grates in the middle of the asphalt expanse. Not that this image makes sense. Any more than a river that flows both ways makes sense.

On the fourth corner stood Lomer Roy's BP gas station, which was protected by two old, unleashed German shepherds. I had my first job there, tanking gas. Perhaps because I was by then a teenager with other things on my mind, I failed to notice that behind the station Chedoke Creek still ran free. Others have since told me that while I was cleaning windshields and checking oil levels, they were being warned away from the open creek behind the station, and the storm drain that carried it under Mohawk Road and the parking lot across.

It takes the edge off the present sense of placelessness, telling these tales of this intersection, this meeting of ancient paths through the woods. "The Crossroads, Barton Heights." During the War of 1812, the British army was encamped on the corner where the gas station later stood, and that is how Major Richard Friend, the British commander, referred to these corners in his official correspondence. *Barton*, for the township that included the village of Hamilton, which was still, by the end of the war, a year away from being more than a dream on paper. *Heights*, for its location on top of the escarpment.

One block south of the Crossroads, Daniel and I follow the arrows on the map and turn right onto Lotus Avenue, then left after one block onto Caledon and face the long, single-storey, brown brick wall of a school. A police officer is talking to three teens who are sitting on the curb. Immediately south of the school is a sunken schoolyard, a large, grassy tract one metre or more lower than the road. A squared vale. The vale continues south beyond the schoolyard, where it becomes the swale that runs behind the houses, one after the other, up Caledon to Jameston. By now I'm humming. The air is humming. I've seen all this on previous visits, but coming as it does at the tail end of our upstream trek, it now makes a topographical sense it did not earlier. It seems clear that what we are tracking behind those houses is the actual creek bed.

This is deep in the heart of Spring Farm. Marlene thought that the creek's spring was located on Jameston Avenue, where it dips low midway between Upper James and West 5th. Could it be under a manhole cover, or in the backyard of one of the houses that front Jameston?

That is, if it still exists. There must be a spring, since the creek flows continuously and we are not living in a state of constant rainfall. Yet we can't be sure. There is no visible evidence, and the storm grate is silent. We have followed Michael and Peter Hess more than two hundred years upstream to this spot, but still haven't really reached our goal. We can't see or say for certain if or where the water, which drew them here, still springs. That we have found the source.

Edmonton, Alberta, Canada, North America, Earth, Solar System, Milky *Way, Universe.*

There was no point in including a return address in the waterproof package my red plastic canoe carried out of sight in the current of the North Saskatchewan River. There would be no one home to return to, for our family was soon moving far away from where the canoe was launched.

No matter. I clearly saw my canoe on its journey under the

stars and across the prairies, entering Lake Superior, crossing one Great Lake after the other and spanning the gap between the twin halves of the constellation *Canadensis*. It would plunge over and miraculously survive Niagara Falls, then continue across Lake Ontario and down the St. Lawrence River. My imagination could not cope with the expanse of ocean, but I dreamed that eventually, somehow, from somewhere, after long travel, a battered plastic hull would return to me. Or at least send a message.

Return it did, though in its own time and not from a land beyond the ocean. The plastic craft did not even make it as far as the St. Lawrence. After the falls, at the mouth of the Niagara River, it turned left, drawn west by the currents toward the head of Lake Ontario and into Hamilton Harbour. It followed me, pulling up to the wharf by Richard Beasley's trading post in the far corner of the bay at Head-of-the-Lake, where Michael and Peter Hess step out, stretch their legs and start walking.

My youthful geography was lacking all along, however. I mistook water-sheds. The North Saskatchewan flows not into the Great Lakes, but Hudson Bay.

Some hand must have re-crafted the landscape, then, and come to the rescue. Water found a way. The message in the canoe is almost washed out, hard to decipher after all these years. It reads, "the water will carry you there and back. It will carry you two places at once; where you want to go and where you have come from."

What the Earth Enjoys

When Glooskap clambered over rock

did the earth dent

where his foot fell? Did water fill the spot

to over flow?

Did the little creek follow sleepless

the bed his finger traced, and

where was he going

what did he know of our watershed?

There is a rock. It is an outcrop, a ridge, more accurately. A ridge of lime-stone. My creek, if I may call it that, Chedoke Creek, begins at or near the base of this ridge. It begins as water rising from out of the earth.

Spring Farm was established because of that spring, in the shadow of the ridge. Three generations later, in the 1860s, the farming operation had evolved into a quarrying business using the rock of that ridge. The quarrying business lasted for another three generations, until the late 1940s, when the city overtook the country.

I've been running upstream and down long enough now to feel that I owe it to the landscape of the creek's birth to find out more about this rock.

The map blew around in his hands, although he had unfurled it only far enough to show what he wanted me to see. It was windy and colder than the day had promised. Bob Geddes and I were parked on Limeridge Road, which used to be a major east-west arterial road but had each of its major intersections pre-empted by the on-/off-ramps of the highway that now runs beside and parallel to it across the Mountain, usurping its role. We were standing beside our cars in a cul-de-sac where Limeridge now dead-ends just a few metres shy of Upper James.

Bob is a geologist, and his maps were of rock formations. We were standing, he showed me, on a transition zone from one rock formation to another. The limestone ridge beneath our feet is an outcrop of that transition.

Limeridge Road is named after the limestone ridge on top of which it rides. Duh. I'd always wondered. But it's like the creeks of the Chedoke watershed all over again: How could anyone guess it exists? The landscape here has been altered, and gives no clue. There is no ridge or rock to be seen.

A young man took it upon himself recently to spend every weekend for two years of his life walking up and down a major downtown Hamilton street. Before so many of the city's industries closed down or left town, the street was a thriving commercial thoroughfare. Dereliction is the current theme. The young man walked up and down the street, praying, weaving a stream of words between all that he could see and all that he could not.

At the event where he spoke, he did not tell us the content of his prayer. In the old days, it would have been directed toward the salvation of souls for Jesus, and it would not likely have been a silent prayer. These days, in the world of evangelical Christianity to which he belongs, the approach is more empathetic and neighbourly. I hoped that it included the buildings that he walked by, the brick walls, the windows and doors, and the plywood that covered many of the windows and doors. I hope he also put in a good word for the sidewalk and the storm grate at the curb; for the tree of heaven and Manitoba maple growing between sidewalk and wall, that never-say-die flora in an area of town which is now about as far from the Garden as you can get.

The prayer-walker walked Barton Street in the north end of the city. Miles away from the Crossroads, Barton Heights. Miles from the garden that was Spring Farm. Same township, different time span, different worlds.

David Hess looked up from the plow one day and wondered if maybe it would be easier to make a living out of the rock outcrop just beyond the spring than from the earth he was cultivating.

That's not quite how it happened.

David Hess was a grandson of Michael and Charity Hess, the first settlers. He continued the farming tradition and family line at Spring Farm, and was still on the job in 1862 when his wife, Elizabeth, passed away. He was sixty years old. He retired from the plow and went to live with his son in Bay City, Michigan, but not before persuading his daughter Charity and her husband, Dan Gallagher, to take his place.

A year later, David was back from Michigan with a new wife, Sarah. For a time the couple lived with Charity and Dan, but then David decided to build a house for himself and Sarah. He chose a location a little downstream of the frame farmhouse, closer to the Caledonia Road, which is what Upper James Street was called at the time, and built a cottage made of stone.

Cottage is a bit of a misnomer. It was a storey and a half, with high ceilings, and, like many such cottages, it was surprisingly commodious. For the stone, he went no farther than the limestone outcrop he had been eyeing earlier. His son-in-law, Dan, helped out. When the neighbours saw what they were doing, they began to ask for rock and lime for their own building projects.

Thus was Barton Lime Works born.

That is to say, the terms of agreement were changed.

When David and Dan launched the quarry and lime works, the pact was altered between Spring Farm and the landscape. The relationship between people and place that Vanessa Watts writes about in her essay was rejigged. The pact had, of course, already been altered three generations earlier, when the land was first surveyed and parcelled out in free grants and the farm was established.

Historically, this pact is altered on a regular basis, at our whim. The earth, our partner, has very little say in the matter.

Or, it seems as though the earth has little to say about it.

Bob and I returned to the car for a brief tour of Spring Farm. He pointed out the limestone rocks used as garden edging in people's front yards and along their driveways – chips off the block the houses are built upon. Their basements must have been blasted out. We turned north from Limeridge Road onto Hawkridge Avenue and stopped at Quarry Court; names resonant with the area's past life. On the corner, Bob saw a rock not placed by gardener, but part of the outcrop itself bulging from the lawn like the low, grey hump of a surfacing whale.

He took out his canvas knapsack, a worn and battered veteran of many field trips, removed from it a geologist's hammer and wondered aloud what the neighbours would say if he did what he then proceeded to do. As I stood watch, he delivered the single, quick blow that freed a flat, five-centimetre chunk of rock and sent small shards to ting against the car as they flew into the grass.

He picked the piece up and held it out to me. The inside surface twinkled like grains of sugar. I checked the fender as we got back in. Undamaged.

Glooskap, in various spellings, is a name found in stories told by various Indigenous peoples from the northeastern Atlantic area of North America. In some of these stories the bearer of this name has a creator role. Local landforms are described as his direct handiwork, or the effect of his running or jumping. In some stories, he has a less creative, more destructive brother.

I wrote the lines about Glooskap clambering over rock at a time when I was trying to integrate my religion, the stories and belief structure within which I was raised, with something a little more local. To see if they could fit amicably together. If I could write something that felt consonant with both.

That's as far as I got. Maybe my better angel was making me feel a little uneasy about the project, and not just because Glooskap is not local to Head-of-the-Lake.

At the back of my mind was something the Ontario poet

James Reaney had once written, to the effect that as immigrants, refugees and settlers, we will never feel at home here, will never truly understand it, until we make peace with its gods.[5] Or did he say spirits? Gods seems almost too lofty; spirits too ephemeral or undefined. I knew what he meant, though. It has something, everything to do with the living reality of this physical place, with what is invisible but definite and present, with how the invisible is made manifest in the world before our eyes, and how we engage in conversation with it.

One of the other maps Bob showed me was a topographical map from 1960 of west Hamilton Mountain produced by the Army Survey Establishment, RCE (Royal Canadian Engineers). It was based on aerial photography, he said, which would later have been confirmed on foot. By a walker.

The map shows brown contour lines at ten-foot intervals. Creeks are thin blue lines. Major roadways are red lines, solid when paved, dashed when unpaved. The undeveloped areas are white, punctuated with small black squares, which indicate houses and other buildings. Outposts of the urban march. In the developed, pink zones, such individual structures are no longer noted, having been swallowed whole. Orchards are orderly rows of small green circles, while clustered black dots indicate abandoned quarries, like that of Barton Lime Works.

The already developed neighbourhoods between the escarpment face and Mohawk Road are a uniform pink. The territory of Spring Farm, bounded by Mohawk Road, Upper James Street, Limeridge Road and West 5th Street, is pied: some areas are urban pink, others remain an undeveloped white. One pink patch is at the corner of Mohawk Road, where the Millionaire Drive Inn and my mother's Dominion grocery store are already or soon will be standing. The other patch is centred midway, on Jameston Avenue. I now think of Jameston as the driveway into Spring Farm, which morphed into the driveway for Barton Lime Works, which morphed into the main cross street of the neighbourhood. All of them leading down into the dip where the spring was located.

Spring Farm on this map also has a name – a new name: Yeoville.

Yeoville?

In the world of Place-Thought, figures like Glooskap, or Sky Woman, or the left- and right-handed twins always at odds with each other are not mythological beings. They are not inventions, or metaphors, or acts of imagination that are meant to explain the world. They are not characters in a story to tell children. They are real.

I have never considered the people who populate the stories that I grew up with as anything less than real, or their stories anything other than true, so let it cut both ways. Let the prayer-walking wilderness guide of my upbringing in the Judeo-Christian tradition, the one who made the map I go by, also be the walking, talking, earthling brother to, if he will, Glooskap.

Let them parley.

On the topographical map from 1960 that Bob showed me, there is no thin blue line to indicate Chedoke Creek. Not anywhere on Hamilton Mountain. Not even a short filament. Nothing in Yeoville. Nothing on the other side of Upper James behind the gas station where I worked. Nothing in the Colquhoun Park just above the falls.

Yet I have my sources. Living, breathing human beings who

lived there at that time, and who testify to an open creek in these and other locations.

Perhaps my witnesses misremember, or a simple time lag is at play. But not even a thin thread of blue shows in Chedoke Gorge, where the creek has always and forever run free. Bob said that these aerial maps were later confirmed on foot. Perhaps the walker spent the day sitting on a park bench instead.

Or, the lines on the map are not factual, but represent a different kind of truth. They show the world as it was perceived at that time. And the truth is that by 1960 Chedoke Creek was no longer in anyone's perception of west Hamilton Mountain.

The Barton Lime Works kiln produced the lime, which, added to sand and water, could be used as mortar in building a stone wall.

To produce lime, the rock was broken by sledgehammer into pieces small enough to handle, which were then loaded onto a horse-drawn cart, which was driven to the top ledge of the quarry and dumped into the kiln below. It could take all day to fill the kiln. The kiln was fired underneath with hardwood to a high temperature and kept going for two or three days. The lime was then drawn off the bottom of the kiln and allowed to cool, after which it could be loaded onto carts and hauled away to construction sites.

Much of the product from Barton Lime Works was delivered down the escarpment into the city. Stone houses and buildings were fairly common to Hamilton from the 1830s onward, though most of the earlier quarries were in the escarpment face. Brick largely replaced stone starting in the 1860s, and it was easier to handle. But by then basements were being dug under most homes for the first time and the stone was used for foundation walls.

Bob Geddes was surprised about the lime production at this

outcrop because it is dolostone, perfect for using in the steel-making process, which was why the industry grew in Hamilton, but not so great for mortar.

Captain Cornelius Park is a continuation of the limestone ridge that runs west of Spring Farm, on the other side of West 5th Street. At the park's north end is a small wooded area of winding, well-beaten paths and campfire pits among large rocks. It's the only place where the outcrop is still visible in its natural state. I wanted to show Bob. It's also just up the street from where I lived as a teenager.

The snow started as soon as we got out of the car: blowing pellets aspiring to be hail. We climbed up from the road into the woods to a clearing where a line of large rocks bulged from the earth like an exposed and weathered foundation wall. We walked up to a single, solitary, rounded and pocked rock that sat nearby. It was the living embodiment of experience and deep reflection on being out in the elements for eons, on simply *being*. A meditative and wise-looking rock.

Bob had his hammer in his hand. He said, "I hate to do this." And then he did it. He gave a quick rap to a small nose that protruded from the face of the rock, and the nose dropped to the ground. He picked it up and put it to his own nose, then handed it to me. It smelled like oil, bitumen. The compressed lives of flora and fauna from under a long-lost sea.

The nose spent a few days in the coffee cup holder of the car before coming into my house and finding a place on a shelf. Not your typical memento. It is neither the crushed stone of the quarry, nor a beach stone rounded to perfection by water, but a bit of both. I think of it as a small obstruction, the kind that causes the creek to ripple.

The stone nose on the shelf also reminds me that Church of the Resurrection has finally responded. They are the church at the corner of Mohawk Road and Garth Street, from whose parking lot a path leads to the Hess Family Burial Ground. I sent them a note through their website asking about the cemetery and relating the story from one of the cemetery's neighbours, in which large numbers of headstones were removed during the housing survey construction in the 1960s and are stored in the church basement.

The present rector, Michael Deed, apologized for the delayed response. It's Lent, a busy time. He has been at Resurrection only a few years, and the church is actually a relatively new congregation, the amalgamation of two others: St. Bartholomew and St. Timothy. But he knows some of the earlier history. Accounts differ, he says. Most of the headstones were destroyed by vandalism, it seems. They have four or five in their basement that have been reassembled and rubbings taken for the sake of preservation.

When the word *reassembled* appeared in his email, I imagined nose-shaped triangles of broken headstone.

Michael also says that the cemetery doesn't belong to the church. They maintain it, but it belongs to the city.

Barton Lime Works was renamed Gallagher Brothers, Lime and Stone, when David Hess finally retired for good and Dan Gallagher (and his brothers) took over the business. Later still, it became Gallagher Lime and Stone Company Limited.

It was a horse-intensive business. Horses pulled dump carts of broken stone from quarry to kiln and wagons of cordwood to the oven. When the wood-burning days were over, they pulled water wagons from the spring to supply the steam boiler. Horses delivered the stone for foundation and house walls in freight wagons to job sites. They delivered finished lime in carts.

To accommodate the many horses, a new and bigger barn was built in 1909.

To facilitate watering the horses and cleaning the stalls, the barn was built directly over the creek, which ran in a stone sluiceway through the middle of the barn, covered with large stone slabs.

This was Chedoke Creek's first experience of being buried. Buried alive.

Standing in the woods of Captain Cornelius Park, which was never my childhood or teenage haunt, the only real memory I have of being here is someone else's story. My mother's. With her sister.

Both sisters were emigrants from the Netherlands in the 1950s: my mother, Anna, came to Canada with husband and two young daughters; her sister, Tina, went to California with husband to start a family. They saw each other only a few times over the next three decades. Tina and Uncle Jack visited Hamilton once in the 1970s, her last trip before early onset dementia made travel impossible. She was in a nursing home within a few years.

That final visit was when both sisters were in their late fifties. I am already older than her, as is Bob, my brother of the rock hammer, standing in these same woods.

Mom took Tina for a walk through Captain Cornelius Park only to discover that the woods, small as they are, and the paths, wide and well travelled as they are, caused increasing confusion and distress for her sister. "Where are we? How do we get home from here?" The weathered and wise rocks were no help. Mom directed them back onto the paved sidewalks again as soon as possible. Her sister's condition was an upsetting revelation. She was also afraid for herself and her own future.

From this limestone outcrop, the creek flows north, under our suburbs, and goes over the falls and runs into the marsh and onward into the bay, the lake, and somewhere on this journey it ascends into heaven and eventually falls back to earth as it has over ages and through eons. It finds a way, however long the cycle may take, and I would like to enter this reciprocity, this back and forth in time. I would like these worn and wise rocks, which were no help then to the two sisters, to spring a few words now from under the earth, words for the afflicted one and words for the one who was later afflicted, words that are current and clear and offer a cup of cold water, dipped from within that eternal circle, and that lend an arm for their careful walk home.

Religious practice takes many forms. Silent praying for the street, as that young walker was doing, seems positive. He is, I hope, as I imagine the creek to be, that is, fully engaged in the life of the place, the better and the worse.

It's all in the prepositions. You pray *to* God and the Great Spirit, invoke Glooskap and Jesus, the Queen of Heaven and Sky Woman, on behalf of, *for* the people who live on Barton Street in the north end of the city of Hamilton, which lost so much of its industrial base over the past forty years that unemployment and its attendant woes now go two or more generations deep. And if you're me, you also pray for the challenged buildings that line the street and side streets, and for cracks in the sidewalk; for the non-human world, which is also suffering.

Or, you drop the prepositions and simply pray God the Great Spirit, trees, the sky, the brick, the person walking their dog around the park, the trash can. Call them forward into your attention. Attend, with no agenda or plan. No ask. It's not your business to decide what others may need or want.

It's about the invisible world, and acknowledging all that you cannot see behind and within the walls the material world places before you.

Being there and speaking. Speaking up. Speaking under and over, around and through whatever lies before your path in this landscape. Like a creek.

Things made and unmade.

Six years after it was built, the new horse barn blew down in a windstorm.

This time round, the Gallaghers decided not to rebuild. The lime kiln and quarrying business was suddenly, by 1915, no longer so horse-intensive. The company had begun to grow a fleet of motor trucks.

The rock slabs were removed. The creek was daylighted.

The creek follows sleepless the bed that Glooskap's finger traced.

The stretch between its source and its waterfall is a bit drowsy, though, dropping only one hundred feet (thirty metres) over the four kilometres. It's relatively flat, which I think is the reason it was so difficult to see the path of the creek bed in the current suburban landscape.

From the top of the falls and through the gorge to the face of the escarpment, the creek drops almost twice as far: one hundred and seventy feet (fifty metres) in half a kilometre. In the three kilometres from escarpment to mouth is another two hundred feet (sixty metres.) Four hundred and seventy feet (one hundred and forty metres) over a little under eight kilometres.

I'm reading these numbers (and converting them to metric) by counting the light brown contour lines on Bob's map like the rings of a tree.

The creek is a living memento of the ice age that strung those lines and shaped the landscape. It's our ongoing connection to a frozen time, when the circle turned even more slowly. How many seasons did that cold, white blanket cover our world? How thick was its pile? When the wind at last blew warm again and Glooskap pulled the cover back; when he ran to break the ice jam that dammed the St. Lawrence River valley, and the freed waters rushed headlong to the sea,

how long did those meltwaters run,
how many years' work was it
do you think? for the torrents to carve
to such rough beauty
our rock-hewn and tumbled gorge?
And after long retreat, when the glaciers dripped dry,
where exactly did he place his hand
to touch the earth
with fingertip
or tap the outcrop, to free the waters
hidden in stone?

Yes, where in Yeoville is that spring now?

Our next-door neighbours, here below the escarpment, in this postglacial era, where we live on a sandbar overlooking Chedoke Valley and the mouth of the creek, are moving south. After almost forty years in one place, they have purchased a house with less stairs to climb, up on the Mountain, where Liz said she never wanted to live. The common urban/suburban tension that exists in many cities is compounded in this city by the Mountain, the wall of the Niagara Escarpment, which physically separates the two.

We will miss them. They have been good neighbours.

Liz told us the new address, so Mary and I went to check it out. The neighbourhood is called Greeningdon, and is next door to Yeoville. The house stands at the corner of Ridge and Limeridge Road, at the cut-off end of Limeridge that lies east of Upper James, opposite to where Bob Geddes and I met, and in the shadow of the limestone outcrop. Liz and Brian have moved from creek mouth to very near the source. Go figure. The creek is playing us all.

The highway and its ramps are audible but not visible on the other side of a high, grassy berm that is the limestone ridge in green disguise. Looking at it, I remembered that I once bagged potatoes up there. A barn stood at that corner of Limeridge Road and Upper James, and as a teenager I spent time in the barn helping Harry Groenewegen bag potatoes into five- and ten-pound paper sacks with metal twist-ties at their tops.

The barn was doomed. Everyone in our Dutch-immigrant community at the time knew that. The city had expropriated the property for the cross-Mountain highway that was being planned, and everyone figured that Cornelius Groenewegen was now a rich man. He is not the Cornelius of Captain Cornelius Park. Captain John Richard Cornelius was a WWI veteran and well-known amateur athletics coach. Harry was related to Cornelius G. through a cousin. A couple of years older than me, he was paying for his education by bagging the potatoes that Cornelius purchased by the truckload from the Ontario Food Terminal in Toronto and sold to local retailers under his own name.

Thinking that I might untuck a corner of the blanket of urbanization and look underneath to locate the Chedoke spring in the memory of someone who had worked close by, I called Harry on the phone.

No such luck. He remembered crossing the feeder creek that went under Mohawk Road at Millbank Road on his way home from school, where I delivered newspapers a few years later, and he remembered skating home on Mohawk when the creek flooded the road, but that was a kilometre farther west. Nothing in Yeoville.

He said I should talk to his uncle Ted.

The city block that started off as "wilderness" and was Spring Farm for almost two centuries is currently a neighbourhood of approximately four hundred buildings, excluding garden sheds. Twenty-five of those buildings are commercial, of which a few are strip malls with more than one business in them. There are three schools and three churches and eighteen backyard swimming pools, plus more than a few acres of pavement in the form of roadways, driveways and parking lots.

The names of the major streets that define the block make sense. In their distinct geographical, historical or cartographical way, they connect with where they are. Limeridge Road rides on top of a limestone outcrop. West 5th Street is the fifth street west of Upper James. Mohawk Road was originally the Mohawk Trail. Upper James (formerly the Caledonia Road, and the Port Dover Road, towns that are the road's destination) is *Upper* because it is the Mountain continuation of downtown's James Street, and *James* because Nathanial Hughson, an early city landowner, named it after his son.

Yeoville is another story.

Sir James Yeo, a British fleet commander during the War of 1812, was famous for blockading the American fleet in Sackets Harbor on Lake Ontario for months, and capturing Oswego. He also outmanoeuvred and escaped American ships that chased him from York (Toronto) to Head-of-the-Lake (Hamilton) in what has become known as the Burlington Races.

Someone in the right place at the right time must have been a history buff. Water is the only connection between Sir James and this landlocked neighbourhood: the water of Chedoke Creek that replenishes the marsh that feeds into the bay that joins the lake that kept our hero's boat afloat.

The water that flows invisibly beneath its streets.

Yeoville's first permanent building, the Spring Farm farmhouse built by Michael and Charity Hess when they arrived in 1789, burned to the ground in 1877. Its residents at the time, Dan and Charity Gallagher, built a new brick house for themselves on the other side of the Caledonia Road. It had gotten too busy, noisy and dusty living so near the quarry.

No one died in the blaze, but the family archives, including the original deed to the property, were lost, and the fire also destroyed many of the records that David Hess, who was by then seventy-five years old, kept as treasurer for Barton Stone Church.

David had helped to build Barton Stone sometime in the 1830s, when he and his fellow Presbyterians decided to go it alone after meeting with the Episcopalians in a Union Church on the Mohawk Trail for a number of years. They chose a location on the Caledonia Road one concession south of Spring Farm, on a corner of the farm owned by Jacob Hess, the eldest son of Michael and Charity, at what is now called Stonechurch Road.

A stone house to give the spirit and soul of the community

a home. To acknowledge that we are in this all together. A place where Jesus and Glooskap and other locals of their lineage can take off their shoes, sing a few songs, pray a few prayers to the Great Spirit God and take a break from the heavy lifting.

Water should spring from such a building, as from a rock.

All that I know of religion is the healing of ourselves and others and of the earth. All that I know of paradise is the dream of this particular place on earth restored.

Barton Stone Church still stands one hundred and eighty years later.

Driving up, I try to picture David Hess's Sunday morning journey with his wife in horse and cart, watching them pull out of the Spring Farm drive at Jameston Avenue and turn right onto the Caledonia Road. Their horse slows down a little as it feels the climb to the top of the limestone ridge, which I really only notice now for the first time. It's not a huge climb, and easily missed in a car. At the crest of the climb, the road crosses a bridge that carries it over the cross-Mountain highway, which is called the Lincoln Alexander Parkway, or locally, the Linc – the nickname of the loved and honoured man it was named after, who never in his long life drove a car. The highway exposes and provides an inside view of the limestone outcrop, which was blasted through for its passage.

Topping the rise, David gently applies the brakes to his horse cart for the slow descent to the next concession road. The limestone outcrop divides the Chedoke from the Red Hill Creek watershed, which drains east over the escarpment at Albion Falls and into Hamilton Harbour. David rides through one of Red Hill's creeks as he nears the corner where the church stands.

The church that he and Elizabeth, and later Sarah, attend is as plain as plain can be. If he and his fellow founding parishioners were aiming for architectural humility, they nailed it. Maybe this was all they could afford: a simple rectangular structure under a sloped roof with a 4/12 pitch, gable end facing the street. The design is reminiscent of a nineteenth-century blacksmith shop or warehouse. The four tall, rectangular window openings on each side are a little more church-like, as are the two front doors set into an arch at the gable end. The doors are painted bright red, as if to blow a kiss to traffic as it blows by on Upper James.

The building's beauty is in its simplicity, the stone longevity, the weathered persistence. Across Stonechurch Road is a car dealership, all sheet glass and stucco, rising from an asphalt plain, fronted with bold, illuminated signage. The diagonal corner houses a drive-through Tim Horton's coffee shop, with more of the same. The asphalt plains of both businesses cant toward storm grates that are the only nod to the creek that still flows under and around them. Opposite the church on Upper James, a few modest homes from the 1930s stand bravely behind a billboard that promises more than their sorry frames can possibly deliver. Something bigger and better. They won't be there much longer.

You'd never suspect that this corner had a story at all if the church and those houses weren't still standing. There is something almost more appealing in the dereliction of Barton Street downtown than in yet another commercial outcrop in the empire of the automobile, which is a form of dereliction to the notion of place, where the goal seems to be to prove that there is no story. It's all present tense. We need a prayer-walker here too. Maybe there is one already. More than one. Maybe those ruby red lips of Barton Stone Church part, and prayer-walkers emerge to speak out and up for all this bought and fraught territory.

The church cemetery extends south from the building, separated from Upper James by a row of mature maples. I found David Hess, who lies buried with his first wife, Elizabeth, at the farthest reach of the cemetery, but searched in vain for Sarah. David and Elizabeth lie together at a point where the land dips down to a second creek. This and the creek that flows under and around the car dealership and coffee shop meet a little farther east. The landscape here remembers the creek in this remnant swale, which continues behind the doomed houses on the other side of Upper James. A number of other members of the Hess family lie near David and Elizabeth, their headstones newer replacements for the originals, which must have been worn down to illegibility by the weather, like most of the stones that surround them. Their names and dates are sharp and clear. Someone is keeping this first family's memory alive. Judging by their numbers, and the care, it makes me wonder what the story is with the so-called Hess Family Burial Ground behind Church of the Resurrection.

Among the four hundred buildings in the neighbourhood of Yeoville are the grade school that I attended for one year when our family first arrived (although the old school was torn down and rebuilt), the high school I attended for four years (although the gym and classroom addition constructed between my grade nine and ten years was demolished, and the remaining building repurposed, and the school itself has moved elsewhere) and the church that our family attended (still standing, and much added to, front and back).

I spent many hours and some very formative years at that southeast corner of Mohawk Road and West 5th, which, although I was not aware at the time, also happens to be the northwest corner of Spring Farm. Is this where my pact with this landscape started?

The gods of the natural and built geography took me in. They were being kind to a displaced-feeling boy who hated to leave the prairies.

The young Peter Hess posed a few questions for his father, Michael, on their journey to Head-of-the-Lake.

"Father, why are we leaving home? What practical, political or religious cause has broken your loyalty to our familiar and loved landscape? Is this migration necessary or voluntary, a star in your eye that we must follow?"

If a migration is voluntary rather than forced, if it is the following of a star, then how does a person grow that loyalty back again in the new land? Does it come naturally? Is it like a coat you put on? What does the receiving landscape have to say in the matter, and why, in your faithlessness to the abandoned landscape, would it trust you?

I pluck an apple from a tree in a corner of the Hess family orchard that is now the parking lot of the church and school of my youth, toss it up and down in my hand and take a bite, thinking of half-worms.

As an apple, Peter Hess rolled a little farther from the tree than did his brothers, Jacob and Samuel. They became farmers, while Peter joined forces with his friend John Mills and, in 1816, purchased land below the Mountain near the centre of the brand new townsite from Margaret Rousseaux, the widow of Jean-Baptiste, a mill owner in Ancaster. Within a decade the townsite took the name of another recent migrant-settler, George Hamilton. The buying and selling, and reselling, and mortgaging, and remortgaging of granted lands in the new settlement had become a major preoccupation almost from day one.

Because not many people were living here yet, there were limited marriage choices, and the branches of family trees became heavily intertwined. John Mills married Peter's sister Christina, and they built a house at what is now the corner of King and Queen Streets in downtown Hamilton. The Scottish Rite mansion that stands there today is a descendant of the original dwelling. Peter himself married Sarah Beasley, the daughter of Richard and Henrietta Beasley, two of the earliest settlers below the Mountain, and built a house on the other side of the street and a short block east of John and Christina, between what is now Hess and Caroline Streets. Street names and resident names also intertwined in settler times. Caroline was their first daughter. The families lived near enough to each other that they could call back and forth through the trees between their front porches.

At the time, the land the two young men purchased was more rural than urban. It stretched from the escarpment to the bay, and from Bay Street to Locke. A sizable swath. Both families kept large garden properties around their homes and grew vegetables, herbs and fruit trees, and raised animals. Both landowners were savvy. They knew which way the wind was blowing and the creek was flowing. They donated road allowances across their properties, and began to sell individual lots in the growing town.

What his family's migration from their Pennsylvania home seems to have taught Peter is that all land is real estate.

The rector at Church of the Resurrection, Michael Deed, passed my name onto Clare Stewart, who is the church historian.

She has responded to say that indeed, Michael is correct, rubbings were made of the three headstones in their basement, by her daughter, in fact, and though the rubbings have been lost, their texts, as much as was legible, were written down:

In memory MI AEL HESS who No 1804 age ears;
In memory of CH TY HESS who died 1808 6 years
In memory of JACOB HESS Who Born Nov 15th 1766
And Departed this Life Oct 7th 1823 Aged 57 years 2
months and 7 days

So it is a family cemetery – for the first settlers and their eldest son, at least.

Clare says that there were never very many stones in the cemetery, contra the retired high-school teacher who told Daniel and me that the cemetery was full of headstones. I wonder if the three remaining headstones are the top portions of the three broken teeth I found hidden in the bushes at the top of the knoll. The third would have belonged to Jacob. "Can I see them?" I ask.

They're in storage and not available for viewing. But she'll see what she can do.

My loyalty is to the landscape. I almost said first loyalty.

I am on the side of this landscape, not as real estate, but as somewhere to live and breathe and have my being. Home to body, soul and spirit, without which there is no place to build the house we call home. There is no story, and we do not exist.

I am on the side of this landscape that lies before me, that has taken me, a latecomer, a migrant, in. I make claims for it. On the map of how I perceive this place, a star is attached to the tip of Princess Point, beside the waters of Cootes Paradise, just past the mouth of Chedoke Creek. In the bottom corner of the map, the key will inform you that the star indicates this spot is the centre of the universe.

If I had stayed living in Edmonton, my loyalty to its prairie landscape would undoubtedly have grown deeper. It still has a strong pull. Perhaps all I am talking about is a simple loyalty to the earth itself. To the clay from which we are made in the origin story that I grew up with. To the pawful of soil in the Sky Woman story that the otter retrieved from water's bottom and brought to the surface. Live anywhere long enough and the landscape will claim your loyalty, your heart. But it's more complicated in an urban area, where it can take the better part of a lifetime for the shape and lay of the land to break through the distractions that are thrown up to hide it from you: the engineering, the built and unbuilt and rebuilt geography, the abuse and dereliction. It can take a lifetime to discover and uncover the physical reality and beauty of the place. If it happens at all. If you're even inclined that way. There are natural barriers that a child born to reason and commerce instinctively throws up against that loyalty. Throws up for self-protection, partly, against any feeling for the place. Or throws up so that they can continue to do whatever they want to do, without conscience, from peeing in the stream to tearing down a stone barn to driving a highway through a valley.

Count yourself lucky if the land breaks through at all. Lucky, and burdened.

Harry Groenewegen, my fellow potato bagger in the barn on Limeridge Road, gave me Ted's phone number, and I called, but not long into our conversation Ted confessed that he was probably not much help, and suggested that I talk instead to his sister Jackie Hogeterp.

My own Dutch apple has fallen and rolled some distance from the community tree, but suddenly I felt those roots begin to quiver again.

The original, post–World War II immigrant families who came to this particular corner of the country have been planting and tending their own family orchards for several generations now. Their children have had children, who have had children. The branchings have interwoven and grown thick. Banquet halls are rented for family gatherings and reunions. It's like the nineteenth century all over again. Everyone is connected to everyone else.

In this case, Jackie is the mother of a good friend of mine, Paul.

The Ontario Genealogical Society confirms that there were, indeed, only eight known burials in the Hess Family Burial Ground. Clare has sent me a copy of their report.

In addition to Charity and Michael Hess, and their son Jacob, there is Jane Snider, the wife of a local farmer, who died in 1820 at the age of thirty-nine and lies under the one remaining headstone in the cemetery.

Also buried on the knoll is the unnamed infant child of a couple named Servos, who died in 1821.

And three more, whose graves were marked M.R., S.R. and M.R. Two of these are the children of Margaret Rousseaux, who lies beside them. That woman knew some grief in her life.

Jacob Hess was the last to be buried there, in1823 at the age of fifty-seven. By that time, two other cemeteries had opened, one beside the Union Church on the Mohawk Trail, the other by Barton Stone Church. These became the preferred eternal resting places for the area.

Naming the cemetery after the Hess Family came about after 1897, when an article in the *Hamilton Spectator* on local abandoned landmarks included a sketch *In the Deserted Graveyard*, which showed the two standing stones of Michael and Charity, and their son Jacob's stone leaning away from a tree just outside the fallen picket fence, their names all clearly visible. Such is the power of art. Until that time, formally or informally, it had been called the Henry Smith Farm Cemetery, after the first owners of the property, and the Terryberry Cemetery after the second owners. Terryberry, like the library. We're coming full circle.

The Terryberrys, by the way, also came here from Pennsylvania in the late 1700s with an Anglicized German name, Dürrenberger.

I thanked Clare for sending me a copy of the report, and told her I'd still like to see those stones in their basement.

Jackie Hogeterp is a wealth of story.

Her parents, Cornelius and Adrianna, purchased the barn where Harry and I bagged potatoes five years after emigrating from the Netherlands. It was their home as well as a place of business. They purchased it from Dr. Bethune, who lived in a stone house that faced Upper James. The barn was part of the original farm property, which, like Spring Farm, had evolved in the nineteenth century from farm into a quarry-and-lime operation because of the limestone ridge. Three walls of the barn's foundation were dug out of the ridge itself, providing cool temperatures that proved perfect for storing potatoes. Cornelius ran his business from there, while the family lived upstairs in the granary.

The couple arrived in Canada in 1947 aboard *The Waterman*,

the first of the Dutch immigrant ships, with five children including Ted, who was six years old, and Jackie, who was three. Their family grew to nine children, all sharing a two-bedroom apartment that took up less than half the space in the granary. Their water came from a pump in the front yard of Dr. Bethune's house, carried in buckets, and later from a city water truck.

Cornelius built a larger apartment for the family in the empty half of the granary, and started renting out the first apartment to other new immigrants. Jackie said that over the years many newcomers came and went through the barn, and through its unusual, square silo, which her father also made into a small apartment.

Jackie was eight when they moved into the barn, and attended grade three at Barton School, south on the highway from where they lived. The next year, she moved schools, and waited as her father watched for gaps in the traffic and walked her across the highway. She carried on walking along Limeridge Road and down West 5th Street to attend grade four at the new Christian school close to Mohawk Road – a school that her father had helped to start. He also helped to start the church on Mohawk Road nearby. The school and church are two-thirds of the reason my father chose to move our family into the neighbouring new subdivision when we came to Hamilton almost fifteen years later. The high school hadn't yet started, but would soon, in the basement of the church. A three-acre Dutch corner of Spring Farm.

This is what Dutch Calvinist immigrants did when they arrived in Canada after World War II: started their own churches and their own school system. Their convictions led them to act in ways similar to Mennonite communities. In religious temperament, they are a little like Mennonites and a little like Presbyterians, who are Scottish Calvinists. While only twenty percent of the Dutch who immigrated were of our particular bent, they were the only Dutch people that I knew existed in Canada.

Cornelius bagged potatoes and some onions and apples in the basement level of the barn, and started a business collecting damaged wooden food crates from grocery stores in Toronto. He repaired the crates and sold them to local farmers. Jackie's summer job was working for her father: bagging vegetables, repairing crates or going to the farmers' market, which was also on his work itinerary. Like many immigrants, he did whatever his hands could find to put bread on the table.

After Dr. Bethune passed away the Groenewegens purchased the Bethune house. Adrianna decided the stone house was too fancy for their family, so they rented it out. The property between house and barn was now available, and when her dad set the first pallet of stacked, broken crates on the ground, it marked the beginning of the end, Jackie said, for the park-like setting of lawn and garden and a laneway lined with cherry trees. Other pallets, rows of pallets, soon followed. Goats were brought in to keep the grass down.

I'm beginning to wonder if it isn't always the beginning of the end.

Jackie lived at home in the barn until she got married in 1966, the year running water also arrived. After she got an indoor tap, her mother loved to wash dishes. Jackie and her husband moved west, but on trips home their children could see and visit where their mother grew up. The barn came down in the 1970s, although the highway was not completed until 1997. I later saw a photo and realized that this was no ordinary barn but one made of cut stone. It was a large and almost regal-looking version of

Barton Stone Church, with the square silo as its steeple. So gorgeous in its own way that it makes you want to cry.

The lost parkland was upsetting enough, but the ruination spread. "It's sad there now," Jackie says, of the whole area around where the barn once stood. Her son, Paul, puts it more strongly. He has taken his own children to see where their grandmother grew up, though there is nothing left to see but cars and the on-/off-ramps of the highway. He calls it a "godforsaken wasteland," and says that there is no place on earth he would rather not be.

This song is the song of the exiled psalmist who goes on singing the Lord's *song in a strange land, reminding the people of an order of value that persists.*[6]

It is totally in keeping with the one-sided, routinely altered pact the refugees, migrants and settlers made with their new homeland that a four-lane highway would be allowed to obliterate the stone house and its stone barn – and anything else in its path. That a quarry could end the cycle of planting and harvest. That the crops and orchards could fence out the wide-ranging Indigenous hunters and travellers who do not *own* the land, for that is impossible to them, but who find themselves caught in this surveyed net, which makes them trespassers in a landscape that moves ever further out of recognition, so that the young First Nation's girl that Marlene Gallagher's mother and Gail Dawson's father see as children while skating Chedoke Creek stands out as unusual and remarkable. Out of place. Exotic.

It's not always the beginning of the end. It's simply the next event in another chapter in the book of what the earth puts up with, how it suffers us and how others suffer, as we move from one use of it to the next.

Jackie Hogeterp's family story unfortunately brought me no nearer to the Chedoke Creek spring. She had no memory of it, or of any creek.

I contacted her younger brother Earl who reported back to me on a small, weekend reunion recently, where the siblings talked about the creek, racking their brains to shake out a memory of some kind for me to chew on. They are universally agreed that there was no creek, though a large area behind their grade school on West 5th often flooded, if that was any help. Since the area is within the city block that comprised Spring Farm, I'll take it. Chedoke Creek wetland.

Anyway, the creek didn't interest them nearly as much as the lime kiln. It was still standing then. Two storeys tall. Abandoned, derelict, with the oven's yawning mouth at ground level. He remembered the kiln as dangerous, scary and irresistible. They all did.

I should write about that, they thought.

There is a visible world and an invisible world.

The visible world before our eyes is the shapely and beautiful, bruised and derelict surface of the invisible world. Scratch the surface and everything goes deeper. Behind and beyond it is everything that makes the surface possible, gives it shape and body, allows it to be and to function: from the two-by-six studs behind the drywall to the world inside the leaf, the rock, the human body. The visible world is the physical manifestation of an invisible world far more vast and complex. Everything goes on forever.

Stories, events and memory are in the invisible world. What happened two minutes ago shares space there with Michael and Peter Hess as they find the spring, and with the first human walking on the moon.

Things unmade remain standing in the invisible world. In my book, this does not mean that the invisible world is all in your head. It is real. An immigrant potato barn still stands at the corner of Limeridge and Upper James. A family of three headstones lean atop a knoll. A spring boils from the ground, or splashes from a crack in a limestone outcrop.

The invisible world is where all these places and things and events go that are now lost or out of sight. Just because you can't see them or they happened in the past doesn't mean that they no longer exist, or that they have no agency.

In my book, the Stewards of Cootes Watershed live up to the terms of a two-way pact when their volunteers haul out garbage from the gorges, ravines, valleys and waters of the creeks that empty into Cootes Paradise – of which there are more than twenty. Deep in East Spencer Creek gorge, close to Tew Falls, they recently discovered a car, a rusted and decomposed fifty-year-old Chrysler. Piece by piece, they have been cutting it up with a portable cutting torch and hauling it out. I witnessed a troop of stewards one Sunday afternoon exiting the gorge with long pieces of metal sticking out of their backpacks. As winter approaches, only the transmission and the engine block remain.

The terms of this pact are not written down, but its fine print is in the leaves and rocks, and in the water.

The two scientists from Redeemer University, Edward Berkelaar and Darren Brouwer, together with Janelle and other students, are living up to the terms by annually keeping tabs on the condition of the creek waters and offering the results to the city.

There are other people and groups equally involved in

reading the fine print. Tys and the RBG. Hamilton Conservation Authority. The Naturalists' Club. Environment Hamilton. And more.

Hamilton Water and Wastewater continues to play catch-up with the illegal hookups on the Mountain that send sewage into the storm drains and contaminate Chedoke Creek and Cootes Paradise. It's a slow and methodical process, sending letters, making phone calls, following up, knocking door to door, offering a free fix and no recrimination.

Urban life is complicated. It's impossible to keep up with everything that human beings and citizens, individually, in groups and commercially, contribute to the natural environment. Or what they destroy. Not everyone's first loyalty is to their landscape. Or second. Not everyone cares. One homeowner answered the door to an employee of Hamilton Water & Sewer and said, "I don't recycle and I don't use blue boxes and I don't compost and I don't give a shit about what goes into Chedoke Creek or Cootes Paradise. Now bugger off."

In my book, the unceasing prayer-walkers of any or no religious persuasion are pact-keepers too.

As a way to personalize and perhaps move things along with my request to see the broken Hess Family Burial Ground headstones that are stored in their basement, Mary and I, along with a visiting friend Lee, attended Church of the Resurrection one Sunday morning. There I met Clare, who has been so helpful, and Mike the priest, who first introduced me to Clare via email. Clare's roots on Hamilton Mountain are seven generations deep, I learned. She is fruit of an orchard planted in eighteenth-century Head-of-the-Lake: United Empire Loyalist, U.E., descended from families who came following the American Revolution.

Her tree branches include many familiar Mountain names. Young. Terryberry. Almas. Shaver. There is even a Hess perched on one. She has a physical investment in the place that I as an immigrant child cannot plumb.

Even before the service began, heads got together and plans were in place to pull the stones out of storage within the next month, and I'm invited to the viewing.

We three guests enjoyed the church, the worship and the cheerful, welcoming leaders and parishioners. As we were leaving, I noticed a sign in the front hallway with a statement on it that was neither the usual faith declaration nor a snapshot history of the congregation, but rather a version of a statement that has become familiar at the opening for many public events, including Hamilton's city council meetings. *Church of the Resurrection acknowledges that the land on which we gather is the traditional land of the Haudenosaunee and Anishinaabe . . . We seek a new relationship with the Original Peoples of this land, one based on honour and deep respect.*

You wonder how they as a congregation and we as a culture can act on this desire, since we are not just part of the problem but at its root, even if we are seven generations into feeling and enjoying ownership of the place. The words reminded me of the desire expressed in Vanessa Watts's essay about the agency of the non-human world, a desire among Indigenous people to regain their pre-colonial mind. After all I've been thinking and feeling in relation to Chedoke Creek and its current watershed landscape, I see where they're coming from. Following the service, Mary, Lee and I took the path beyond the parking lot to the cemetery knoll, where I showed them the three broken teeth of headstones in the bushes, although this time out I could locate only two. We walked a few metres and parted the nearby cedar's branches to give greetings to Jane Snider, who still stands upright and intact. With the splash of what may be vandal's paint over her inscription, I get more attached to lonely Jane and her standing stone each time I visit through the branches.

The retired high-school teacher who claimed that the church had allowed trucks to drive through the cemetery during construction of the subdivision after first uprooting hundreds of headstones, many of them Hesses, was in his backyard gardening with his wife. I wanted to question him a little more closely, since I now had a few facts that contradict his story, but the morning was already long gone, we were hungry for lunch and I just wasn't in the mood.

Mood? Historical research should not be subject to mood, should it? This may be the last time I go to that cemetery and see him there. He may die.

I didn't like passing on the opportunity, but decided to have faith in the creek. If what that man has to say is crucial to its story, another opportunity will present itself.

Water is very forgiving. Flexible. It will find a way.

I keep calling it Chedoke, but the creek was originally known as Hess Creek on the Mountain, and later it was called Gallagher's Creek. It was significant enough a stream that in the late 1890s Gallagher's Creek was suggested as a water source for the small community that had grown up around the Caledonia Road near the escarpment edge, an area the newspaper referred to as Duff Town, which had a Chedoke Post Office.

To recap: the name *Chedoke* first appears in a real estate offering. Chedoke House was built in 1836 and still stands at the edge of the escarpment two kilometres west of Upper James, beside the gorge that the creek runs through. When William Scott Burn built it, he wondered in a letter to a friend why others were not taking advantage of the view.

The town of Hamilton below the escarpment and the more rural area above it both lay within Barton Township. As the town grew southward, it annexed more and more of the Mountain until by 1973 the township officially ceased to exist. Nineteenth-century maps show the creek winding across a mountain landscape of sectioned farms and large private lots like a snake in a game of Snakes and Ladders. In my meandering and roundabout quest to follow Michael and Peter Hess and to find the source of the creek, the winning square is Spring Farm. But I keep sliding down the invisible snakes of my own willingness to be distracted. To follow my nose whithersoever it might lead. Or so I thought. Now I wonder if, for reasons of its own, the spring does not want to be found. That it prefers the invisible world.

The stone shards were spread out on a red blanket on the basement floor of the Fireside Room in Church of the Resurrection. We stood looking down at them: Clare Stewart; her husband, John, and son Iain; and Michael Deed, the minister. Michael had gotten to the room ahead of us, removed the pieces from their resting place in the unused hearth and laid them out.

An impossible puzzle, I thought. The pieces were all sizes and jagged shapes. Lettering was clearly visible on some of them, but reassembly into three headstones looked hopeless.

Michael had matched together a number of pieces already, though. He had an eye for subtle differences in the carved lettering. Now he and Iain got down on their hands and knees and went at it. They seemed to share a spatial gift. Fielding advice from above, which they bore gracefully, within half an hour the impossible had happened and the three headstones of Charity, Michael and Jacob Hess were put back together.

If it had not been obvious earlier when we saw how irregular in shape and size the pieces were, it certainly was now. These headstones had not fallen over and broken into pieces in the natural course of earthly events. On each stone a web of lines radiated from a centre where the first blow of a sledgehammer had struck and exploded them. In Michael and Charity's case it was dead centre of their names and explained the missing letters, which were turned to splinters and dust. The violence was perfectly transcribed in the remaining broken stone and missing shards. In anger or malice or for no bloody reason at all, someone had taken down these three who, for almost two hundred years, had stood quietly minding the world from the top of a small hill.

"I just don't understand vandalism," Clare said.

Michael returned from his office with *Mountain News* clippings from the 1980s that he had found in church files about the attack on the headstones. The event predated the formation of the present church and congregation, and is not part of their collective memory.

Our little group parted soon after. Michael had official business to attend to. Clare, John and Iain planned to come back in the afternoon to map and number all the pieces, and return the assembled stones to the hearth in the Fireside Room, separated by sheets of cardboard.

In an article that he wrote for the *Mountain News* in 1976, author David Faux, U.E., lamented the direction that Upper James Street was going. The cultural, not compass direction. Especially south of

Mohawk Road. The Crossroads, Barton Heights was becoming "an ugly neon commercialized jungle," as he called it.

"It takes a sharp eye to see the early homes tucked between the compacted blocks of tasteless modern structures that obliterate the landscape," Faux continued.

You could expect that he, like Clare, might care, given the U.E. behind his name and the seven or more generations of living here that it implies. His article was about Barton Lime Works, and the homes he was talking about were those of the Gallagher and Hess families. He was remembering Spring Farm. Marlene Gallagher showed me the newspaper clipping in her family album on another visit.

David and Sarah Hess's stone cottage still stood at the time, but was demolished less than ten years later, in 1983. Consigned to the invisible world. Maybe that act of vandalism finally shattered the newspaper writer's caring too, and he took a sledgehammer to some old headstones standing on a knoll. I doubt that he did it, but anger and hopelessness can turn a person against the very things they love. If the world is not going to show an iota of respect for the story, why pretend that there is one?

Forty years after David Faux's article, I travel Upper James Street, moving, stopping, waiting at the traffic lights at one major intersection, then another, daydreaming, thinking ahead or staring blankly. He's dead right about the "blocks of tasteless modern structures." Big, flat and square are the favoured features of architectures in the empire of automobile; vast tracts of pavement their preferred landscape. It's not simply a dereliction of the sense of place. There is an anti-beauty at work here that institutionalizes a vandalism of the landscape. Our reward is that we get to live with the ugliness and toxicity we have created. In our one-sided pact with the earth it does, in the end, have a say. We're told not to look elsewhere for heaven because all we know of paradise is already with us here on earth. Perhaps the same applies to hell.

Okay, that may be a little strong. A lot of people love this environment and its functional convenience well enough, thank you very much. I myself remember the Millionaire neon sign fondly. And as a famous man once said, why worry? It will all get washed away. As in the past, economic currents and technological developments will reshape the natural and built geography as surely as they did for Spring Farm, Barton Lime Works and C. Groenewegen Produce Ltd. Defining the landscape from day one as real estate has gotten down into our very soul, with the result that we can never feel so attached to this place for its own sake that we value all that we build upon it, or take care in all that we do to it. We have neither made peace with this land's gods, nor exorcised our own demons. We have never become truly at home.

I wish that I could enjoy the current incarnation of the Crossroads, Barton Heights, otherwise known as the corner of Upper James and Mohawk. That I could enjoy being there, or at the next two concession corners southward: amputated Limeridge, with its highway on-/off-ramps; Stonechurch, with its brave stone church.

I wish that what I see would make me want to pull up stakes and go to live beside a spring that rises from the ground, like a certain German-Pennsylvanian dad and his ten-year-old child.

Mark Bainbridge, from Water and Wastewater Planning and Capital, Public Works Department, City of Hamilton, responded to my email query about locating the source of Chedoke Creek.

Mark was the late addition to our "The Story of the Chedoke Watershed" evening program. At the show, he suggested to the audience that they not replicate the storm drain excursion that I, as the evening's leadoff speaker, had described to them. It's too dangerous, he told them. Water levels can change without warning. He spoke calmly and without judgment. I felt chastened but defensive, since our party of adventurers had been very careful to choose a calm and cloudless day, knowing full well what a little rain could do.

Mark told me that his department has many drawings and maps of the area that I defined for him: the southern half of the neighbourhood of Yeoville between Upper James, Limeridge Road, West 5th Street and Jameston Avenue, including geotechnical investigations and site plans. Other than an underground cliff identified near Quarry Court, he has found no clue as to the location of a spring – although he admitted that his maps of the area are dated 1968 or later, which is after the creek and its spring were buried. The department has nothing dated earlier, but the library downtown should have some.

I discovered (later) that the library has none.

He also showed me a map held in the provincial archives pinpointing all the private wells that existed in Yeoville and the immediate surrounding area before they were capped in the 1950s. There are more than two dozen. Wells are usually capped when an area becomes urbanized and is integrated into a waterworks system.

Mark said that the record shows some of the wells on Jameston Avenue were only a little over a metre deep and characterized as *strong* in their pumping rate. The compression of housing development, with its landfill and road paving, will affect the water table and can push a spring down and bury or, in effect, cap it, redirecting it to join the water table at a lower level. There it can infiltrate the pipes of a storm drain system. In other words, a single spring may disappear and a larger area become the source for the creek.

"Just my thoughts," he says.

Water is flexible. Opportunistic. It will find a way. I know as much from car-pentry work I've performed in my own and other people's houses. Water will find the faulty flashing, a gap in the caulking. It will infiltrate, nose forward and pursue an ever-downward route. You may not know how diligently the little creek has been seeking an outlet to the sea through your living quarters until you wake one night during a rainstorm to the sound of dripping on your bedcovers.

You are made abruptly aware that you are within a river's flow.

If the spring of Spring Farm has disappeared, its waters have not. For all these years the creek has kept flowing and falling over the escarpment edge. Perhaps the wetlands described by the Groenewegen family became the source. Or the spring went into hiding, in sympathy with the buried creek. It does not want to be found. The earth, in its agency, has hidden it. And I wonder now, if I did locate the spring, would my first loyalty allow me to share that information, or prevent it?

Perhaps the important thing is not pinpointing the precise location of a spring or source. The important thing is knowing that it still exists.

Down by the North Saskatchewan riverside where I launched my red plastic canoe, and to which the storm drain sent my clothespin boat, I had a youthful adventure, a life-and-death experience.

One spring afternoon, two friends and I walked into the river valley after school. In the valley was a park, and at the edge of the park was a steep drop-off to the river itself. For some reason I thought it would be a good idea to go down the slope, which was comprised entirely of red-brown mud. It took only three steps in my regulation orange-trimmed black rubber boots, which came halfway up my shins (which all the boys wore, usually with their tops folded down), before I was locked into the mud. I couldn't move, and was already out of reach of my friends. I couldn't move because I was stuck in the mud and because I had looked down and become paralyzed by the sight. Springtime in that northern clime meant water moving en masse. Below me the river was a living, breathing, brown being eating away at the bottom of the slope. Behind me a white torrent was bursting from a large, black cave hole in the concrete wall we had skirted. The storm drain. Somehow, we hadn't noticed it until now. It was feeding the river god, and the river god was insatiable.

I thought I was going to die.

My friends ran up the valley slope to someone's house and called my parents. My dad came, disgruntled by this after-work disturbance, reached out his arm and pulled me up and out of the mud. He drove my friends home, I had a bath and that was that. We ate dinner.

"You should be happy I'm alive!" I wailed.

My older sisters just rolled their eyes.

City of Waterfalls is a civic branding launched by an enthusiastic individual, Chris Ecklund, to promote the dozens of waterfalls that grace the escarpment wall within city limits, and to help change the negative image the city-without-a-star has traditionally had in this country. The effort has paid off. Over the past decade, increased numbers of visitors have come to see the beauty of the cascade, the ribbon and the terraced falls, and the gorges their creeks travel through. It's caught the city off guard.

These tourists and out-of-towners come unaware that the rock of the escarpment face is often loose and easily dislodged and slippery when wet, and that it is not wise to venture too closely to the edges, or to climb up or down. The number of rope rescues of stuck or injured people has risen, and deaths have occurred. Albion Falls, which Red Hill Creek goes over, is a particular problem because road access is from the top and the view from the two lookouts inspires the urge to reach the base of the falls. People clamber down steep, unmarked yet tantalizing paths. People continue to clamber on the steep paths even after the city erects fencing and posts dire safety warnings. City staff were on location one day trying to hatch a plan for clear signage and a map board without, at the same time, uglifying the area. A man seeing the newly erected fencing ran up to them frantically and said, "I'm getting married down there in half an hour!" "Sorry, sir," they told him. "You can't. It's dangerous, and illegal." Two minutes later his bride arrived in a white gown, wearing flip-flops.

Flip-flops. People with such a level of geo-incomprehension need to be protected from themselves. This was what Mark Bainbridge was addressing when he tried to dissuade anyone in the audience from replicating our storm drain venture. I feel chastened all over again.

Over the past fifty years, water has found a way to undermine the channels that carry Chedoke Creek alongside the Chedoke Expressway, where Nigel, Dorna, Mary and I walked on that clear and cloudless day. The slabs of concrete have become the cracked and shifted tectonic plates that we enjoyed balancing on and jumping between. But the provincial Ministry of Transportation has had its eye on nature's inroads all along, and is reasserting control. The channels are, as we speak, being repaired.

Chedoke Expressway has also recently been renamed the Alexander Graham Bell Parkway. No one asked me about that change, and I have an opinion.

The current rehabilitation and fix-it list includes the highway-side slope of the channel where Daniel and I paddled in his red canoe, and where my foot sank in the quicksand of the creek bed. The slope had apparently been leaching contaminant from the old city dump beneath the highway. Not a complete surprise.

My description of both these broken, sylvan, surprise wonderlands, therefore, no longer applies because they no longer exist, except in the invisible world. The invisible world is where a lot of the best stuff seems to be located. These repairs were necessary and had to happen, but the decay of the two-generations-old infrastructure was appealing on a number of levels. Nature had patiently undermined the engineered perfection and begun the long process of reclamation. What the water could not accomplish directly on its own, it did by nurturing the roots of trees and reeds and bushes. Both sites were pleasing to the eye, and less brutal, more humane than when they were new. Chedoke Valley seemed not dead but only sleeping, on the road to revival. You could feel less like an exile there.

You could sing a few bars of the Lord's song there, thankful to be reminded of an order of value that persists in this strange landscape we have created for ourselves.

What is justice for a creek?

Working within the cycle of water, it drains the area of its responsibility, its watershed.

As water falls from the sky, it is absorbed by the earth for the trees and other flora, and for the animals to drink. If the sky is too generous and the earth's requirements are met, the bonus water follows a path of least resistance until it can join a liquid line of its kin already on the move. This is not water looking for trouble. It wants off the Mountain, over the falls, down the hill. It seeks the marsh, the lake, the sea.

Trouble comes looking for it, however, in the form of the hard surfaces that replace the earth's soft acceptance, and in pipes that increase its velocity, and in grates that catch bicycles and shopping carts that impede flow, create dams and flood the neighbourhood. Water takes the rap for trouble caused by others.

The sentence is bigger pipes, higher concrete walls.

Where's the justice?

Locals are not immune when it comes to incidents and accidents at the various waterfalls in Hamilton.

My daughter was landscaping one day at a house whose front yard faced over the two lanes of Scenic Drive and the edge of Chedoke Gorge. She saw a line of teenagers walking along the edge. One by one, they ducked and disappeared down a steep path. She thought nothing of it until some came running back across the road, frantic for a phone (in the days before cellphones) because one of their friends had slipped and fallen. For many months afterward, bouquets of flowers and notes adorned the bridge directly over the falls, in tribute to their friend who had died.

Peter, a friend who grew up in the neighbourhood below

Chedoke Falls, regularly led his young nephews and nieces on hikes up the gorge to the falls. Instead of turning around and leaving the way they came in, he guided these ten- and twelve-year-olds up the same steep path the teenager had fallen from.

A young woman told me about being led by her father up the slope and rocky cliff of Tew Falls when she was ten years old. The cliffs often seem more climbable than they actually are, and her father told her only later, as a young adult, that halfway up he realized how stupid and dangerous the climb was. It would have been even more dangerous to try to turn around and climb back down, so he kept going, leading his child on in full and reverent dread of the worst that could happen. She herself was aware of the danger at the time, yet had complete confidence in her father and followed him without fear.

I've done the same. Almost the same. I led my ten-year-old daughter and her friend up Chedoke Gorge along the creek bed. It's a boulder-strewn and therefore laborious walk, especially for short legs, so on the way back I guided them along a path that clung to the side of the gorge, a little higher than halfway up. At one point in this walk, the path became nothing more than a very narrow ledge cut into the side of a very steep slope, which ended below in large rocks and water. The kids followed me, fearless, chatting back and forth the whole time. They wouldn't have missed it for the world. Yet one slip and, well, that's the nightmare scenario that still replays in my head.

As a fearless and trusting ten-year-old, the young Peter accompanied his father, Michael, into the wilds of Head-of-the-Lake. The story of their journey together may be apocryphal, yet it persisted along the branches of the family tree for generations. You have to wonder why. Maybe the story is more interesting than the facts. Maybe the story conveys truth the facts obscure.

The facts are that four men of the Hess, Rymal, Kribs and Smith families travel in an advance party to Head-of-the-Lake in 1788 looking for land. They camp under the trees at the intersection of two ancient highways on the top of a low hill overlooking the elbow of a creek. After scouting the area for a few days, each chooses an area of the countryside for themselves and their families to resettle on, and they return home together. The next year, they immigrate with their spouses and children. Fifty people in total. Plus horses and cattle.

At the core of the monumentality and logistics of such a move is a parent leading their child from security and familiarity into the unknown.

"Where are we going, Father?" the child asks. "We'll know when we get there," the father responds.

They do get there, and they do know. After long travel, the pair reach the mouth of a creek, which they chase uphill to a break in a cliff, then trace the creek's thread through narrow rock walls to a falls, find enough footholds in the steep slope and cliff face to blaze a trail to the top of falls, dislodging and watching rocks tumble down as they go, and then meander for a few kilometres through wood and meadow and wetland to *a crystal clear spring . . ., a spring of pure gushing water . . .*

Made it. We're here. This is the spot. Do we still have some of that cheese left?

There is something mythic, heroic and everyday about it. A migration hike with a happy ending. A Sunday afternoon outing that changes your destiny.

Peter Hess is my neighbour. He lives just a few blocks away. The street half a block away is named Peter Street after him. Peter Street begins three blocks east at Hess Street, at a T-intersection with a disused synagogue on one corner, a pharmacy in a house on the other and a mosque across the street. The city block on which his house stands (in the invisible world) now has a row of townhouses in its front yard, facing King Street, most of which have been converted to businesses, while more houses, plus a fourteen-storey-tall apartment building, take up its backyard. In the two hundred years since he chose to invest in this location, this place, Peter's block has become a model of built diversity and urban intensification. It's what he would have wanted. In his will, he bestowed property lots in the neighbourhood to his children and servants, as well as to his wife, to whom he also bequeathed all the fruit from the apple tree of her choice in their orchard.

One Sunday afternoon, he led one of his own children, and a couple of his sister Christina Mills's children, on an excursion to Chedoke Falls. "I did this with your grandpa when I was your age," he told them. They had never met their grandpa Michael. He died when the eldest was only four. Some of the rocks had shifted and new ones had fallen from the cliff face, along with a few trees, but the gorge that Peter led them into looked much the same as it did when he was a child. And I imagine that it wasn't a lot different from how it looked yesterday when I made a similar creek-gorge excursion with a friend and his two young sons. Chedoke Gorge is like an old stone house. It's like its namesake Chedoke House, weathered, but intact and still standing, with different generations of inhabitants passing each other in its hallways and stairwells.

With Gerrit and Reuben and their dad, Marvin, along, I parked the car at the top of Chedoke Avenue and we walked up a path between the houses to the Chedoke Radial Trail. I pointed out the ravine that carries the original, remnant creek behind the houses on Chedoke Avenue, then turned and followed a path on the opposite side of the trail to the large grate and concrete walls of the storm drain at the mouth of the gorge, into which the creek now flows. We waited as five other Sunday walkers and their small white terrier made their way through the hole in the chain-link fence. The city maintains a fence here to keep people out, but *maintains* is the operative word because someone always cuts a hole in it. A sign beside the hole issues the warning: "No Trespassing. Deep Gorge. Unstable/Uneven Ground. Fast-moving Water. Maximum Fine $5000.00." By my rough calculations, in addition to the maximum twenty thousand dollars the city could have collected from our small group, it stood to make at least another two hundred thousand dollars on all the others who went through that same hole. As a mostly law-abiding citizen in an era of heightened concern about falls traffic and accidents, I am not sure where to put my repeat flagrancy. The city may be learning where to put it, though. On a subsequent visit to this hole in the fence, tickets were being handed out by a bylaw officer to offenders like us. For seventy-five dollars. A little less than the maximum.

Stepping stones helped us over the creek as we started toward the base of Lower Chedoke Falls, one hundred metres ahead. Lower Chedoke is a five-metre-high teaser to the main event, which lies another half-kilometre up the gorge. We climbed over and around rocks the size of living-room furniture, and were suddenly hit with the overwhelming smell of death. The boys accused me, but the source was a raccoon, stretched out and decomposing on the top of a rock. The creek swings from one side of the gorge to the other, under banks of red shale that tower to a level as high as Lower Chedoke. Two rocks, each the size of a small car, are positioned at the edge of the falls, one of them precariously so. It's been poised to fall for years. I wonder if Peter Hess & Co. saw the same sight. The water falls into a small pool and basin. We decided to climb the bank to our right, a steep ascent up wet, claylike soil that had a few footholds cut into it, as well as roots and small tree trunks to grab onto. Gerrit and Reuben, ages twelve and fifteen, climbed up as though it was flat ground, then returned to the aid of their leader and guide, who is not quite elderly enough to be their grandpa, and took his water bottle so he would have two free hands. Their father, Marvin, a younger man, though only by a decade, also took the climb in stride.

We came out above the falls on a path that led back down to its lip, where we stepped onto the backs of the two small car-sized rocks. The rocks of the gorge are not rounded by weather or water, but are flat and squared off blocks from a wall; pieces from the harder, higher layers of the escarpment that have been worked free by time and the elements. Some blocks break off and tumble and slide into position in the creek, while others make it no farther than the sloped skirt of rocks at the base of the cliff, where they lie beached at downward trajectories, inching their way to the water's edge, at the mercy of their own inertia. The overall effect is of a big kid who hasn't put away his building blocks. A giant.

Someone more human-sized had also been at play. Two towers of stacked creek rock stood balanced atop one of the larger rocks. Chedoke inukshuks.

The sloped bottom half of the gorge walls is populated with tall, thin trees reaching for the light. Their roots are regularly undermined by the high water that comes rushing through during storms, and they fall on the slope and across the creek in every direction. Gerrit and Reuben used these fallen trees as walkways and bridges, bouncing up and down on one that cantilevered over the creek like a diving board. None of us were jumping into the water, however. I had been careful to relay the facts about Chedoke Creek and the presence of E. coli. We were each wearing closed-toe sandals, and the plan was to wash thoroughly upon our return home.

Hopscotching rock, water and trees, we slowly travelled upstream until Chedoke Falls appeared in the distance, a translucent form of life that stood out through the foliage. The falls are scaled to their surroundings. Impressive without being overwhelming. Welcoming. It draws you toward it. You want to get there and stand in its spray. No wonder people get themselves into trouble trying. Chedoke is one of two falls at the end of the gorge. Denlow Falls becomes visible only as you near the end of the gorge. Much smaller than Chedoke, that day it seemed no more than a leak; water emerging from a storm drain and spreading down a jagged wall of layered rock into a tumble of fallen rocks and trees at the bottom that almost entirely hid its brief journey to the creek. Water, yet again, finding a way. Flexible. Fifty metres south, people were leaning over the concrete bridge rail of Scenic Drive above, watching Chedoke Falls emerge from its storm drain. They could see the water as it swept over the edge, and they could see us too, but the white curtain that was falling before our eyes was invisible to them. We'd earned this view.

The pooled water in the large, hollowed basin looked a pale green. A sign stood cemented to one of the rocks beside the pool. "Danger. Keep Out. Unsafe for Swimming or Drinking. These waters are known to contain high levels of E. coli bacteria and/or sewage which are a risk to your health and safety." Beyond the sign, two male/female couples were in various stages of entering the water. The water was cold, judging by the screams, but the couple in bathing suits managed to get all the way in and to swim around. Then the young man got out and took photos of the young woman lying on her back, floating in the water that would occasionally wash over her face. The other couple, in shorts and T-shirts, entered only as far as their knees. A third couple arrived, stripped down to their bathing suits, posed against the rocks for selfies and asked Gerrit to take a few shots of them, then waded in, floated and swam around.

The boys expressed no interest in entering the water, but Reuben gingerly made his way over the wet rocks to stand behind the falls. Gerrit followed, and the two made a circuit around the pool and basin, bending to accommodate the overhanging rock. The five Sunday walkers who had preceded us through the break in the chain-link fence now arrived with their little white terrier. They must have carried him through certain portions of the gorge, or he was one intrepid little dog. His legs and belly were wet. He lapped the water in the creek.

Chedoke Falls has apparently become a popular spot for everyone and their dog to visit. It was always popular with neighbourhood kids. A swinging rope, now gone, once hung from a tree above the pool. Maybe the kids have gotten wise. Maybe they can read. Meanwhile, stories continue to rise with the mist. A few years ago, the owner of a home at the top of the gorge stepped outside with his morning cup of coffee in hand and heard calls for help from below. A young woman had been abandoned at the base of the falls the previous night by the two men who had led her in, and she was disoriented, cold and afraid. This past summer, a father and mother and their two children became stranded on a rock when a storm broke and the creek suddenly turned into a mad, raging torrent. They were rope-rescued out. As I was pointing out to Marvin the infamous steep path that my friend Peter confessed to leading his nieces and nephews up, the same path that the teenager had fallen from and died, the young couple who had gone into the pool only to their knees began to climb up, on bare feet, their bundled clothing tucked under one arm, the other helping maintain balance as they clambered upward. Before long the two disappeared from sight behind the rock and foliage. Gerrit saw them and immediately wanted to do the same. Youth. They just go. Maybe it was Peter Hess who led his father up that path to the top, rather than the other way around. Michael Hess was in his late forties when they made the journey. With a wife and seven children still at home, and the eighth child at his side, caution may already have taken hold of his soul. Maybe that's why he asked his youngest son along in the first place. To keep him going. Flexible. Their venture into the unknown required the fearlessness and trust of a child. Maybe that's the crux of the story's longevity. Child companion as inner child made visible.

More people arriving at the falls passed us as we headed back down the gorge. A constant, shall we say, stream. Almost all were young, in their twenties and thirties, often in couples, but a few were older parents showing this world to their children, and although everyone was operating within their own cone of silence with regard to strangers along the way, as we crossed paths one of the fathers said to me, "It's beautiful here, isn't it?"

The usual array of plastic bottles and other washed-down debris littered the creek and its banks. A selection of crushed beer cans adorned the top of a large, flat rock. Four or five bath towels clung soddenly to rocks at different points along the way. *Bath towels?* A half-rolled carpet lay wedged between rocks. It was beyond ridiculous. I hoped the Stewards of Cootes

Watershed included this section of Chedoke Creek in their cleanup schedule. Two trees standing beside each other were tagged in spray-can black from ground level to as high as the tagger could reach. Their designs suggested totem poles, which helped take the edge off the desecration. Adding to the overall gorge effect of the day, the creek had a murky look, which is unusual in my experience, and the rocks at the bottom were covered with a wavy green toque of algae. I had been thinking that this stood as one of my least favourite trips up the gorge, one in which the challenge to nature and the stress on running water in an urban setting was on full and sorry display. Yet he called it beautiful.

We spent a few minutes lolling about at the top of Lower Chedoke Falls, where the boys practiced how close they could get to the edge without plunging headlong. Gerrit picked up a rock the size of his own head with the intention of throwing it over, but we looked first to make sure the coast was clear and saw between the rocks blocking our view at least a dozen adults and kids enjoying the falls and basin below. A man and a woman began to make the climb to the top of the falls using the route on the other side of the creek from the one we had taken an hour ago. It's a straight up, steep slope of clay and rubble, and it led them to the ledge that is part of the rock layer that the water falls over. To keep going, they had to negotiate a few steps along the narrowest portion of the ledge, leaning slightly outward over the abyss, while using the crumbly rock layers above them for handholds. It was touch and go, but they succeeded, handing their takeout cups of coffee back and forth to each other as first the one then the other performed this brief, death-defying act, all the while laughing together in fear and surprise.

I invented the takeout coffee cups, but that couple did seem far too unconcerned.

Marvin was already waiting when the couple reached the wider portion of the ledge, with Gerrit and Reuben following right behind. All three easily navigated the route to the other side through that needle's eye of a thrill, which left only me, muttering underneath my old-man scaredy-cat breath, "This is crazy," as I inched forward, holding onto the narrow, Jenga-like blocks of stone that stuck out from the wall.

So it was no surprise at all to me when one of the stones pulled out like a key from a lock. I lost my balance, and plunged into the pool below.

Chedoke Creek is indeed on the list of creeks that the Stewards of Cootes Watershed cares for and attends to. Alan Hansell said that they visit regularly, and choose the same route around Lower Chedoke Falls that we chose going in, rather than the one of our exit, and they have worked out a system that will allow them to carry out the manhole cover they encountered the last time a team went up the gorge. The system involves a pole and two people. Gerrit, Reuben, Marvin and I somehow managed to miss seeing the manhole cover. Alan figures that the only way to get the cover past Lower Chedoke is to throw it over, then fish it out of the pool below. Should be fun. He's looking forward to it. He's sending students from the university who have recently pledged three hours of weekend volunteer cleanup for the ten weeks of the fall and winter terms. Twenty weeks total.

I find this effort and energy inspirational. A balm.

Water is very forgiving, yes, and flexible. It has had to be, considering who and what it is dealing with while sustaining life on earth. There is always something new that even as humble a waterway as Chedoke has to put up with. Most recently, the Combined Sewer Overflow tank in Chedoke Valley, which is supposed to contain raw sewage during big summer rainstorms, malfunctioned, creating a big stinking spill in the creek and bolstering the notion that its name should be Shit rather than Chedoke.

At the same time, assaults on the Chedoke landscape continue. The latest is a plan to landfill more of the valley that has not already been lost to fill and highway in order to allow two residential towers to be built. When it came to a vote, even a recommendation from Hamilton Conservation Authority staff, guardians of the watershed, to not approve the project couldn't compete with the development benefits to the city.

Knowing how things are in the world, from the ongoing willingness to alter and destroy a landscape, to the E. coli in Chedoke Creek, to the microfibres of plastic that now show up in tap water around the world, who isn't overwhelmed? Even my innocent red canoe, launched half a century ago, is part of the problem. By now its plastic has broken down into tiny pieces that will never go away.

You want to ignore or forget, but still be aware and not stick your head in the sand. You try not to abandon hope and shut down, but how much power does one person have over what's happening, to alter the direction in which everything seems to be flowing?

So you bow your head, chant a walking prayer and attend to what's lying right in front of your feet.

Only in my dreams does the stone block dislodge Jenga-like from its slot so that I lose my balance and fall, like water, over the edge and down.

I land face up in a pool at the base of Lower Chedoke Falls, in its gentle spray, then pick myself up and clamber easily back to the top, and push the car-sized rocks over the edge just because I can. When did that happen? I heave the rocks back into place and run up-gorge, hand-tagging the trunk of each tree as I go. At Chedoke Falls, I hoist myself up and crawl, on all fours, into the mouth of the storm drain, standing every now and then to burst the bonds of concrete, just because I can. Once the daylighting begins, it's hard to stop. It's liberating. The culverts are cannelloni. I eat them upstream past Garth and West 5th, across Upper James and past Mohawk, all the way to Jameston Avenue, where the asphalt peels back to finally, finally reveal and free the spring of water to gush forth from the earth, and I bend to drink.

Refreshed by the same . . . *crystal clear spring . . . , a spring of pure gushing water . . . , a stream of very cold, clear water . . .* that Michael and Peter called home, it's time to play again, so I borrow a wooden clothespin from one of the neighbours on Hawkridge, twist it apart and place both pieces in the creek. The twin boats move swiftly through the backyards between the houses on Caledon and Hawkridge Avenues. They are swept through the culvert under Upper James Street, and loll through the marshy area beside the Mountain Arena just north of Hester Street, floating slowly behind the businesses on Upper James. They're neck and neck entering the culvert under Mohawk Road and into the parking lot of the Canadian Tire store. I came through here only a few minutes ago with my upending ways, but since that time a wide swath of pavement has been cleared and walking paths line both sides of the naturalized creek channel. With benches. Mature trees. It's lovely and serene. Where the creek turns west and goes under Upper

James again, all of Richwill Road has been unearthed and people on both sides of the street are sitting on lawn chairs, enjoying the creek that runs through their front yards. At West 5th Street, the creek turns as it enters a brick culvert with an arched roof that is lit naturally from above through a series of vertical shafts with storm grates at street level. This is my kind of infrastructure. The waterway sees daylight again as it turns onto South Bend, where it is edged with plant life and paths and people, and kids at work in the mud and stones. Buchanan Park is a wetland marsh, with tall reeds and lily pads, which the ducks have already found. After a leisurely sail down Bendamere Avenue, the creek turns right at West 23rd and enters its ancient, meadowed glory, flowing through Colquhoun Park, carving a channel through the rock before falling over the edge of the escarpment – where I lose sight of my two boats. I jump down the falls and hop from rock to rock, chasing after them through the gorge, barely managing to keep up, then leap over Lower Chedoke Falls. I am back at the pool where this all began. I am also back in the days where anything is possible. And I'm thinking that the earth enjoys this as much as me. The creek, now with a mind entirely of its own, ignores the bars of the storm drain and heads straight for its old course behind the houses on Chedoke Avenue. It veers slightly west at Aberdeen and under the railway tracks, then shoots to the valley, where the six creeks of the Chedoke watershed are singing loud and strong, having conspired to entirely remove any trace of the limited-access highway, the 403, that so recently dominated its width. With a welcome sigh, their waters kiss Cootes Paradise on the lips, and I scoop up my two wooden boats before they get swept into the bay.

And like my brother and I did in Edmonton when we were

young, I take the clothespin halves and run back to where the stream begins, and do it over again.

I just can't get enough of this place.

Trusting that this meandering line of words would one day find the sea, all I really wanted to do is remember my first loyalty and keep faith with the creek while pursuing and being pursued by it, travelling upstream and down through this landscape of valley and highway, culvert and gorge, stone house and suburban development, pollution and vandalism and beauty. With a spring and a prayer.

Now, considering all that Chedoke and the waters of the world must deal with while continuing to sustain life, I'll have to let that faith keep me.

Acknowledgements

Thanks and acknowledgements go to everyone who, in one way or another, whether they knew it or not, contributed to this project:

Anna Terpstra, Gail Dawson, Marlene Gallagher, Bert Schilstra, Nigel Terpstra, Dorna Ghorashi, Paul Hogeterp, Dirk Windhorst, Peter Tigchelaar, Lil Blume, Madeleine Simpson, Edward Berkelaar, Janelle Vander Hout, Rick Hayes, Daniel Coleman;

Miss Vanden Top, Udo Ehrenberg, Tys Theysmeyer, Alan Hansell, Darren Brouwer, Nicholas Terpstra, Roger Baxter, Chris Banitsiotis, Mike Durst, Graham Cubitt;

Bob Geddes, Harry Groenewegen, Michael Deed, Clare Stewart, John Stewart, Iain Stewart, Ted Groenewegen, Jackie Groenewegen-Hogeterp, Nelly Groenewegen-Schuurman, Earl Groenewegen, Mark Bainbridge, Peter & Laura Enneson, Katie Mayne, Marvin Oldejans, Gerrit Oldejans, Reuben Oldejans, Lee Hardy, Al Ernest.

Acknowledgement and thanks also to the Ontario Arts Council Writers' Reserve for its financial and moral support; *Geez Magazine* and its editor, Aiden Enns, for publishing early effort, "Watersheds I have known" (Spring 2016, no. 41); and to *Hamilton Arts & Letters* for publishing earlier versions of the first two sections of this book, in issues eight.2 and ten.2. "Citizen Geography" (published as "Daylighting Chedoke") won the *HA&L* Short Works Prize for Non-fiction in 2016.

Notes

[1] Thomas King, *The Truth About Stories: A Native Narrative* (Toronto: House of Anansi Press, 2003), 2.

[2] Rick Nestler's "The River that Flows Both Ways" (1980) is a song written from within the movement begun in the 1960s to clean and protect the Hudson River and wetlands through advocacy and public education.

[3] *Paddle to the Sea* is a National Film Board of Canada film directed by Bill Mason in 1966 that I probably watched, but can't recall if it was before or after reading the original story in the book by Holling C. Holling in which a carved wooden canoe sets out on a journey with many adventures from Lake Nipigon in Ontario all the way to France and back again.

[4] *Terrain vague* is the French term for *wasteland* or *vacant lot*: urban space that may once have been used or built upon but is now abandoned. In landscape architecture and urban studies, these have become theoretical playgrounds.

[5] I have not been able to locate the essay or quote from James Reaney about Indigenous gods or spirits, and neither have the dedicated friends and relatives who maintain his website. But he said it, I swear. It was in an essay that I read in an anthology in the library in Port Elgin, Ontario, one rainy September day in 1985, when we were on vacation with our two small children.

[6] "This song is the song . . ." is from a review by W.F. Blissett of Kathleen Henderson Staudt's *At the Turn of a Civilization: David Jones and Modern Poetics* (Ann Arbor: University of Michigan Press, 1994) that appeared in *University of Toronto Quarterly* 66, no. 2 (1997): 479–85. This quote has the added benefit of allowing me to repeat the name of the Anglo-Welsh poet (*In Parenthesis, The Anathemata*), painter and essayist (*Epoch and Artist*) David Jones (1895–1974). My lines on pages 69 and 82 (*When Glooskap clambered* . . . and . . . *how long did those meltwaters run* . . .) are thinly disguised homages to his poem "The Sleeping Lord."

Bibliography

Gallagher, John. *The Hess Family of Barton*. Hamilton: John R. Gallagher, 1984.

Holling, Holling C. *Paddle-to-the-Sea*. Boston: Houghton Mifflin, 1941.

King, Thomas. *The Truth About Stories: A Native Narrative*. Toronto: House of Anansi Press, 2003.

Moore, W.F. *Indian Place Names in the Province of Ontario*. Toronto: Macmillan, 1930.

Seavey, J.R. "In the Deserted Graveyard." In *Pen and Pencil Sketches of Wentworth Landmarks*, by Mrs. Dick-Lauder, Mrs. Carr, R.K. Kernighan (The Khan), J.E. Wodell, J.W Stead, J. McMonies et al. Hamilton: Spectator Printing Company Limited, 1897.

Watts, Vanessa. "Indigenous place-thought & agency amongst humans and non-humans (First Woman and Sky Woman go on a European world tour!)." *Decolonization: Indigeneity, Education & Society* 2, no. 1 (2013): 20–34.

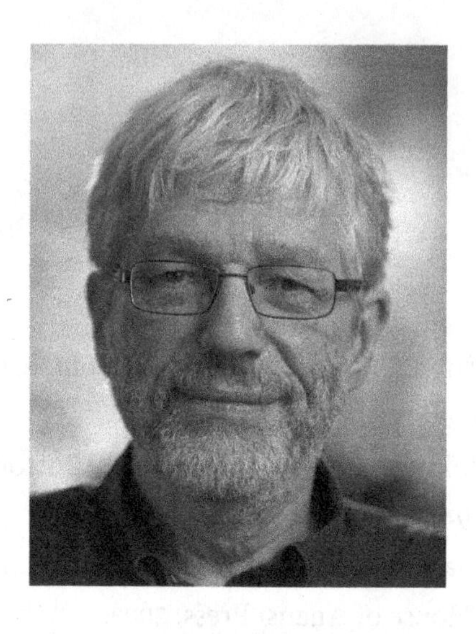

John Terpstra is the author of ten books of poetry and four books of non-fiction. He often plays in that zone where human beings interact with nature – nature in the city, not the country. The nature he gravitates toward is one that has some experience of us, has had to live with us and our demands, and is no longer pure or whole or perfect, but still somehow manages to be itself – maybe even more than when it was "wild." He is interested in how natural geography and built geography integrate and relate to each other, and in how history is simultaneous with now. *Daylighting Chedoke* is a companion book to his two earlier books about Hamilton as a living, breathing geographical location, *Falling into Place* and *The House with the Parapet Wall.*

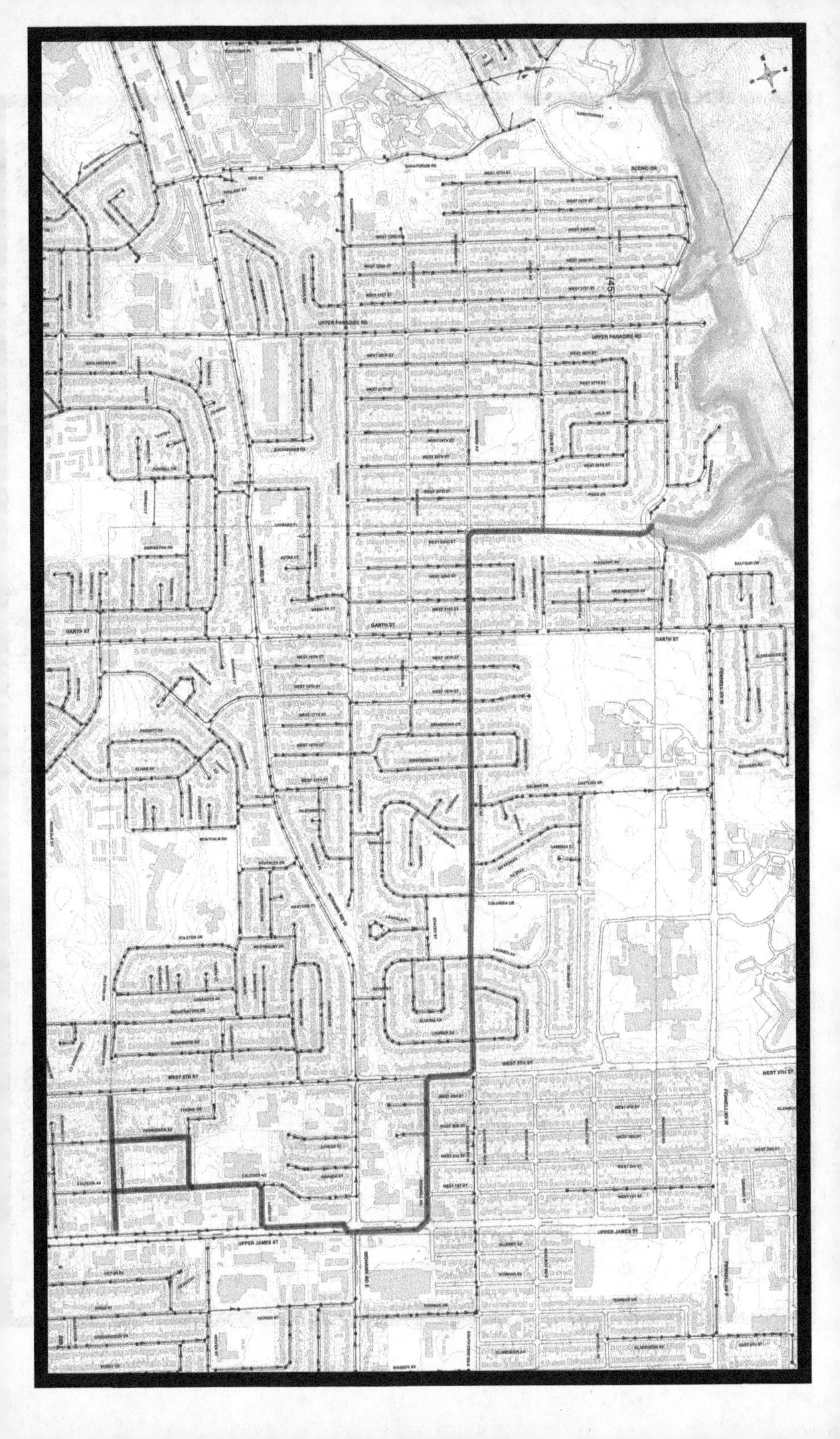
SCENIC DR
UPPER PARADISE RD
GARTH ST
WEST 5TH ST
UPPER JAMES ST

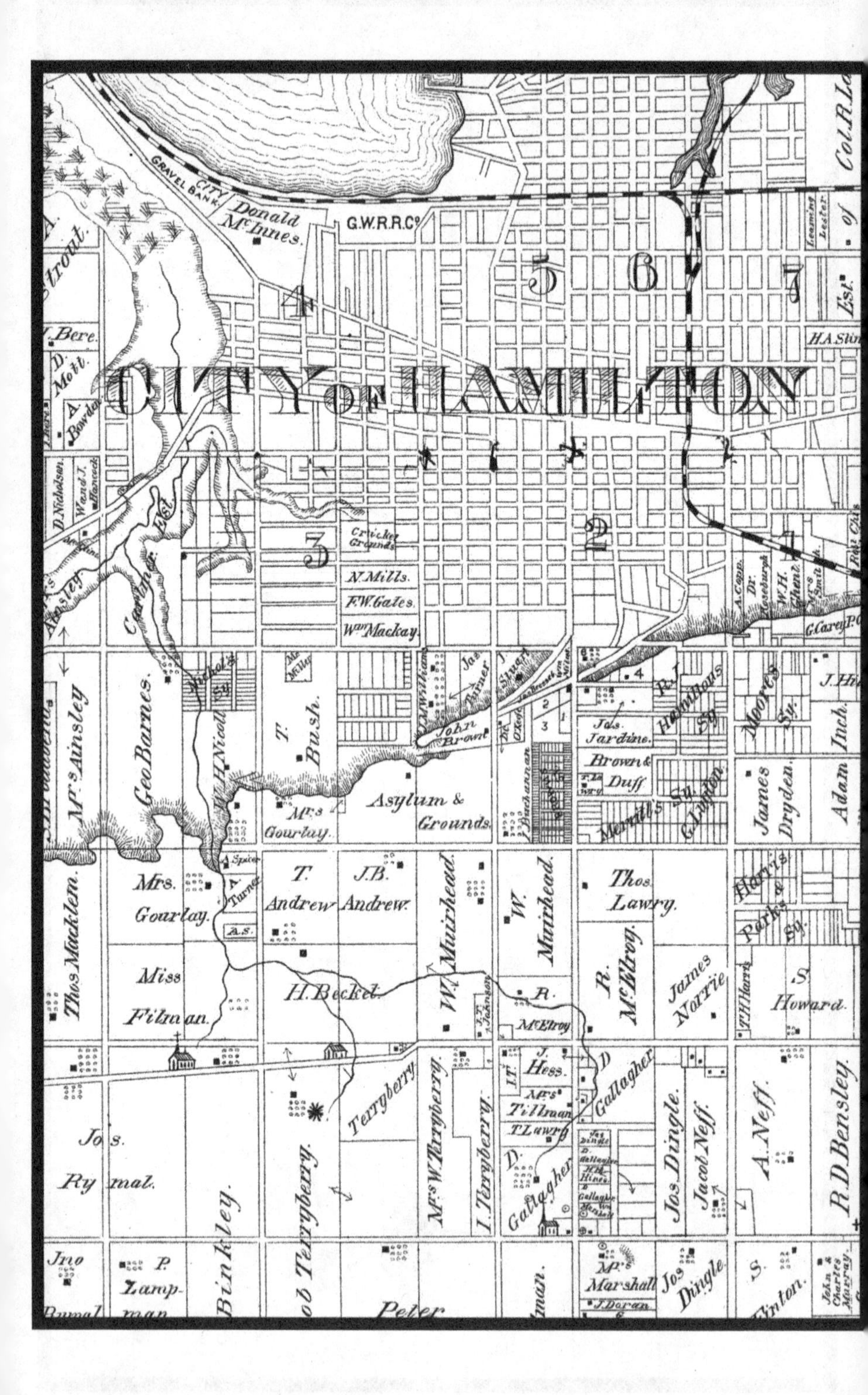

CITY OF HAMILTON
CITY GRAVEL BANK
Donald McInnes.
G.W.R.R.Co
Cricket Grounds
N. Mills.
F.W. Gates.
Wm Mackay.
Asylum & Grounds
T. Bush.
Mrs Gourlay.
Geo Barnes.
Mrs Ainsley
Ainsley
Nichol's Sq.
John Brown
Jas Jardine.
Brown & Duff.
Hamilton's Sq.
Merritt's Sq.
Moore's Sq.
James Dryden.
Adam Inch.
Thos Macklem.
Mrs. Gourlay.
Miss Filman.
T. Andrew
J.B. Andrew.
H. Beckel.
W. Muirhead.
Thos Lawry.
R. McElroy.
James Norrie
S. Howard.
Harris Parks Sq.
J. Hess.
Mrs Tillman
T. Lawry
D. Gallagher
Jos. Rymal.
Binkley.
Terrybery
Mrs W. Terrybery.
I. Terrybery.
Jos Dingle.
Jacob Neff.
A. Neff.
R.D. Bensley.
Jno Rymal
P. Lampman
Peter
Mrs Marshall
J. Doran
S. Vinton.
John & Charles Murray.
G. Carey P.O.
A. Bowden
D. Mott.
D. Nicholson
Wm and J. Hancock
H.A. Slin
Col. R.
Est.